The Winners in Life's Race

This edition published 2025
by Living Book Press
Copyright © Living Book Press, 2025

ISBN: 978-1-76153-879-7 (hardcover)
        978-1-76153-880-3 (softcover)

First published in 1882.

This edition is based on the 1883 printing by D. Appleton and Company.

A catalogue record for this
book is available from the
NATIONAL LIBRARY OF AUSTRALIA
National Library of Australia

The Winners in Life's Race

or the Great Backboned Family

by

Arabella B Buckley

LIVING BOOK
PRESS

Contents

1. THE THRESHOLD OF BACKBONED LIFE..............................5

2. HOW THE QUAINT OLD FISHES OF ANCIENT TIMES HAVE LIVED ON INTO OUR DAY...........................23

3. THE BONY FISH, AND HOW THEY HAVE SPREAD OVER SEA AND LAKE AND RIVER..............................45

4. HOW THE BACKBONED ANIMALS PASS FROM WATER-BREATHING TO AIR-BREATHING, AND FIND THEIR WAY OUT UPON THE LAND. 71

5. THE COLD-BLOODED AIR-BREATHERS OF THE GLOBE IN TIMES BOTH PAST AND PRESENT. 89

6. THE FEATHERED CONQUERORS OF THE AIR (1)..........122

7. THE FEATHERED CONQUERORS OF THE AIR (2)......... 150

8. FROM THE LOWER AND SMALLER MILK-GIVERS WHICH FIND SAFETY IN CONCEALMENT, TO THE INTELLIGENT APES AND MONKEYS. 202

9. THE LARGE MILK-GIVERS WHICH HAVE CONQUERED THE WORLD BY STRENGTH AND INTELLIGENCE........245

10. HOW THE BACKBONED ANIMALS HAVE RETURNED TO THE WATER, AND LARGE MILK-GIVERS IMITATE THE FISH. 284

11. A BIRD'S-EYE VIEW OF THE RISE AND PROGRESS OF BACKBONED LIFE. 315

PREFACE.

Although the present volume, as giving an account of the *vertebrate* animals, is a natural sequel to, and completion of, my former book, *Life and her Children*, which treated of *invertebrates*, yet it is a more independent work, both in plan and execution, than I had at first contemplated.

This arises from the nature of the subject. The structure and habits of the lower forms of life are sufficiently simple to be treated almost without reference to geological history. When, however, I began to sketch out the lives and structure of the vertebrate animals, which are so closely interlinked one with another and yet so sharply separated into groups, I soon found that I must carry my readers into the past in order to give any intelligible account of the present.

I have therefore endeavoured to describe graphically the early history of the backboned animals, so far as it is yet known to us, keeping strictly to such broad facts as ought in these days to be familiar to every child and ordinarily well-educated person, if they are to have any true conception of Natural History. At the same time I have dwelt as fully as space would allow, upon the lives of such modern animals as best illustrate the present divisions of the vertebrates upon the earth; my object being rather to follow the tide of life, and sketch in broad outline how structure and habit have gone hand-in-hand in filling every available space with living beings, than to multiply descriptions of the various species. If my younger readers will try and become familiar with the types selected, either alive in zoological gardens or preserved in good museums, they will, I hope, acquire a very fair idea of the main branches of the Backboned Family.[1]

In order to treat so vast a subject simply and within narrow limits, it has often been necessary to pass lightly over new and startling facts. I trust, however, it will not be inferred that such passages have been lightly or carelessly written, for in all cases I have sought, and most gratefully acknowledge, the assistance of some of our best authorities; and I have endeavoured that what little is said upon difficult subjects shall be a true foundation for wider knowledge in the future.

Among the many friends who have rendered me valuable assistance, I cannot sufficiently express my obligations to Professor W. Kitchen Parker for his unwearying kindness in explaining obscure points of anatomical structure, and to my friends Mr. Alfred R. Wallace, Professor A. C. Haddon of Dublin, and Mr. Garnett of the British Museum, for constant suggestion and encouragement. I am also indebted to Mr. J. P. Anderson of the British Museum for aid in the arrangement of the Index.

The geological restorations given as picture-headings (some of which are here attempted, I believe, for the first time) have been most carefully considered, though the exact forms of such strange and extinct animals must necessarily be somewhat conjectural. My thanks are due to the artist, Mr. Carreras, jun., for the patience and care with which he has followed my instructions regarding them, and also to Mr. Smit for his masterly execution of the frontispiece.[2]

---

1   Almost every animal mentioned in this book is to be found alive in the London Zoological Gardens, or stuffed in the British Museum.

2   The Figures in the text, which, with exception of about twenty, have all been

I have been asked why, in this and the former work, I have not given genealogical tables to help the reader to follow the relations of the various groups. My reason is, that it is impossible to construct tables of this kind without giving a false idea of the fixity of natural divisions and of the extent of our knowledge. To men of science, who know how provisional such tables are, they have a certain value, but they would be positively harmful in a work of this kind, which will have fully accomplished its purpose if it only awakens in young minds a sense of the wonderful interweaving of life upon the earth, and a desire to trace out the ever-continuous action of the great Creator in the development of living beings.

Arabella B. Buckley.
LONDON, September 1882.

drawn expressly for this book, are the work of the above-mentioned artists, together with Mr. Coombe and Miss Suft.

THE GREAT BACKBONED FAMILY

CHAPTER I.

THE THRESHOLD OF BACKBONED LIFE.

LIFE, life, everywhere life! This was the cry with which we began our history of the lowest forms of Life's children, and although we did not then pass on to the higher animals, is it not true that before we reached the end we were overwhelmed with the innumerable forms of living beings? The microscopic lime and flint builders, the spreading sponges, the hydras, anemones, corals, and jelly-fish filled the waters; the star-fish, sea-urchins, crabs, and lobsters crowded the shores; the oysters, whelks, and periwinkles, with their hundreds of companions, struggled for their existence between the tides; while in the open sea thousands of crustaceans and molluscs, with cuttle-fish and terribly-armed calamaries, roamed in search of food. Upon the land the snails and slugs devoured the green foliage, while the vast army of insects filled every nook and cranny in the water, on the land, or in the air. Yes! even among these lower forms we found creatures enough to stock the world over and over again with abundant life, so that even if the octopus had remained the monarch of the sea, and the tiny ant the most intelligent ruler on the land, there would have been no barren space, no uninhabited tracts, except those burning deserts and frozen peaks where life can scarcely exist.

Yet though the world might have been full of these creatures, they would not have been able to make the fullest use of it, for all animal life would have been comparatively insignificant and feeble, each creature moving within a very narrow range, and having but small powers of enjoyment or activity. With the exception of the insects, by far the greater number would, during their whole lives, never wander more than a few yards from one spot, while, though the locust and the butterfly make long journeys, yet the bees and beetles, dragon-flies and ants, would not cross many miles of ground in several generations.

What a curious world that would have been in which the stag-beetle and the atlas-moth could boast of being the largest land animals, except where perhaps some monster land snail might bear them company; while cuttle-fish and calamaries would have been the rulers of the sea, and the crabs and lobsters of the shores! A strangely silent world too. The grasshopper's chirp as he rubbed his wings together, the hum of the bee, the click of the sharp jaws of the grub of the stag-beetle, eating away the trunk of some old oak tree, would have been among the loudest sounds to be heard; and though there would have been plenty of marvellous beauty among the metallic-winged beetles, the butterflies, and the delicate forms of the sea, yet amid all this lovely life we should seek in vain for any intelligent faces,—for what expression could there be in the fixed and many-windowed eye of the ant or beetle, or in the stony face of the crab?

These lower forms, however, were not destined to have all the world to themselves, for in ages, so long ago that we cannot reckon them, another division of Life's children had begun to exist which possessed advantages giving it the power to press forward far beyond the star-fish, the octopus, or the insect. This was the Backboned division, to which belong the fish of our seas and rivers; the frogs and toads, snakes, lizards, crocodiles, and tortoises; the birds of all kinds and sizes; the kangaroos; the rats, pigs, elephants, lions, whales, seals, and monkeys.

Is it possible, then, that all these widely different crea-tures, which are fitted to live not only in all parts of the land, but also in the air above, and the seas and rivers below, and which are, in fact, all those popularly known as "ani-mals," only form one division out of seven in the real animal kingdom?

Can it be true that while the chalk-builders have one division all to themselves, the sponges forming a transition group, the lasso-throwers another division, the prickly-skinned animals a third, the mollusca a fourth, the worms a fifth, and the insects a sixth, yet the innumerable kinds of birds and beasts, reptiles and fishes, are all sufficiently alike to be included in one single division—the seventh? It seems at first as if this arrangement must be unequal and unnatural; but let us go back for a moment to the beginning, and we shall see that it is not only true, but that quite a new interest attaches to the higher animals when we learn how wonderfully life has built up so many different forms upon one simple plan.

Starting, then, with the first glimmerings of life, we find the minute lime and flint builders, without any parts, making the utmost of their little lives, filling the depths of the sea, and wandering in pools and puddles on the land; acting, in fact, as scavengers for such matter as is left them by other animals. But here their power ends; to take a higher stand in life a more complicated creature is needed, and the sponge-animal, with its two kinds of cells and its numerous eggs, is the next step leading on to the curious division of lasso-throwers. These, in their turn, do their utmost to spread and vary in a hundred different ways. Possessed of a good stomach, of nerves, muscles, powerful weapons, and means for producing eggs and young ones, they fill the waters as hydras, sea-firs, jelly-fish, anemones, and corals. But here they too find their limit, and, without advancing any far-ther, continue to flourish in their lowly fashion. Meanwhile the tide of life is flowing on in two other channels, striving ever onwards and upwards. On the one hand, the walking star-fish and sea-urchin push forward into active life under the sea, forming, with their relations, a strange and motley group, but one which could scarcely be moulded into higher and more intelligent beings. On the other hand, the oyster and his comrades, with their curious mantle-working secret protect their soft body within by a shelly covering, and by degrees we arrive at the large army of mollusca, headed by the intelligent cuttle-fish. And here this division too ceases to advance. The soft body in its shelly home does not lend itself to wide and great changes, and it was left for other channels to carry farther the swelling tide of life. These take their rise in the lowly, insignificant division of the worms, which may, perhaps, have had something to do with the earliest forms even of the star-fish and mollusca, but which soon shot upwards, on the one hand along a line of its own, while, on the other, we have seen[3] how, in its many-ringed segments, each bearing its leg-like bristles and its line of nerve-telegraph, the worm foreshadowed the insects and crustacea, or the *jointed-footed* animals of sea and land, forming the sixth division.

Here surely at last we must have reached animals which will answer any purposes life can wish to fulfil. We find among them numberless different forms, spreading far and wide through the water and over the land, and it would seem as if the sturdy crab and fighting lobster need fear no rival in the sea, while the intelligent bee and ant were equal to any emergency on dry ground. But here the tide of life met with another check. It must be remembered that the jointed-footed animals, whether belonging to land or water, carry their solid part or skeleton *outside* them; their body itself is soft, and cased in armour which has to be cast off and formed afresh from time to time as they grow. For this reason they are like men in armour, heavily weighted as soon as they grow to any size, while the body within cannot become so firmly and well knit together as if all the parts, hard and soft, were able to grow and enlarge in common. And so we find that large-sized armour-covered animals, such as gigantic crabs and lobsters, are lumbering unwieldy creatures, in spite of their strength, while the nimble intelligent insects, such as the ant and bee, are comparatively small and delicate.

It would be curious to try and guess what might have happened if the ant could have grown as large as man, and built houses and cities, and wandered over wide spaces instead of being restricted to her ant-hills for a home, and few acres for her kingdom; but she too has found the limit of her powers in the impossibility of becoming a large and powerful creature. Thus it remained for Life to find yet another channel to reach its highest point, by devising a plan of structure in which the solid skeleton should be—not a burden for the soft body to carry, as in the sea-urchins, snails, insects, and crabs—but an actual support to the whole creature, growing with it and forming a framework for all its different parts.

This plan is that of the backboned animals. They alone, of all Life's children, have a skeleton within their bodies embedded in the muscular flesh, and formed, not of mere hardened, dead matter, but of bones which have blood-vessels and nerves running through them, so that they grow as the body grows, and strengthen with its strength. This is a very different thing from a mere outer casing round a soft body, for it is clear that an animal with a living growing skeleton can go on increasing in size and strength, and its framework will grow *with* the limbs in any direction most useful to it.

Here, then, we have one of the secrets why the backboned animals have been able to press forward and vary in so many different ways; and especially useful to them has been that gristly cord stretching along the back, which by degrees has become hardened and jointed, so as to form that wonderful piece of mechanism, the *backbone.*

Look at any active fish darting through the water by

sharp strokes of its tail,—watch the curved form of a snake as it glides through the grass, or the graceful swan bending his neck as he sails over the lake,—and you will see how easily and smoothly the joints of the backbone must move one upon the other. Then turn to the stag, and note how jauntily he carries his heavy antlers; look at the powerful frame of the lion, watch an antelope leap, or a tiger bound against the bars of his cage, and you will acknowledge how powerful this bony column must be which forms the chief support of the body, and carries those massive heads and those strong and lusty limbs.

Nor is it only by its flexibility and strength that this jointed column is such an advantage to its possessors; the backbone has a special part to play as the protector of a most valuable and delicate part of the body. We have already learnt in *Life and her Children* to understand the importance of the nerve-telegraph to animals in the struggle for life. We found its feeble beginnings in the jelly-fish and the star-fish; we saw it spreading out over the body of the snail; we traced it forming a line of knots in the worm, with head-stations round the neck, which became more and more powerful in the intelligent insects. But in all these creatures the stations of nerve-matter from which the nerves run out into the body are merely embedded in the soft flesh, and have no special protection, with the exception of a gristly covering in the cuttle-fish. We ourselves, and other backboned animals, have unprotected nerve-stations like these in the throat, the stomach, and the heart, and cavity of the body. But we have something else besides, for very early in the history of the backboned animals the gristly cord along the back began to form a protecting sheath round the line of nerve-stations stretching from the head to the tail, so that this special nerve-telegraph was safely shut in and protected all along its course.

A careful examination of the backbone of any fish, after the flesh has been cleared off, will show that on the top of each joint (or *vertebra*) of the backbone is a ring or arch of bone; and when all the joints are fastened together, these rings form a hollow tube or canal, in which lies that long line of nervous matter called the *spinal cord*, which thus passes, well protected, all along the body, till, when it reaches the head, it becomes a large mass shut safely in a strong box, the skull, where it forms the brain.

Here, then, besides the unprotected nerve-stations, we have a much more perfect nerve-battery, the spinal cord, carried in a special sheath formed of the arches of the backbone, which is at once strong and yielding, so that the delicate telegraph is safe from all ordinary danger. Now when we remember how important the nerves are,—how they are the very machinery by which intelligence works, so that without them the eye could not see, the ear hear, nor the animal have any knowledge of what is going on around it,—we see at once that here was an additional power which might be most valuable to the backboned division. And so it has proved, for slowly but surely through the different classes of fish, amphibia (frogs and newts), reptiles, birds, and mammalia, this cord, especially that larger portion of it forming the brain, has been increasing in vigour, strength, and activity, till it has become the wonderful instrument of thought in man himself.

We see, then, that our interest in the backboned or *vertebrate* animals will be of a different kind from that which we found in the boneless or *invertebrate* ones. There we watched Life trying different plans, each successful in its way, but none broad enough or pliable enough to produce animals fitted to take the lead all over the world. Now we are going to trace how, from a more promising starting-point, a number of such different forms as fish, reptile, bird, and four-footed beast, have gradually arisen and taken possession of the land, the water, and the air, pressing forward in the race for life far beyond all other divisions of animal life.

On the one hand, these forms are all linked together by the fact that they have a backbone protecting a nerve-battery, and that they have never more than two pair of limbs; while every new discovery shows how closely they are all related to each other. On the other hand, they have made use of this backbone, and the skeleton it carries, in such very different ways that out of the same bones and the same general plan unlike creatures have been built up, such as we should never think of classing together if we did not study their structure.

What the lives of these creatures are, and what they have been in past time, we must now try to understand. And first we shall naturally ask, Where did the backboned animals begin? Where should they begin but in the water, where we found all the other divisions making their first start, where food is so freely brought by passing currents, where movement from place to place is much easier, and where there are no such rapid changes as there are on the land from dry to damp, from heat to cold, or from bright leafy summer, with plenty of food, to cold cheerless winter, when starvation often stares animals in the face?

It is not easy to be sure exactly how the backboned animals began, but the best clue we have to the mystery is found in a little half-transparent creature about two inches long, which is still to be found living upon our coast. This

Fig. 1.

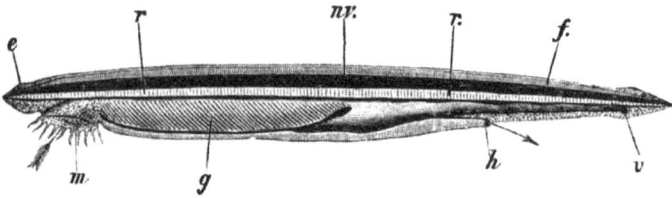

The Lancelet, the lowest known fish-like form.

m, mouth. *e*, eye-spot. *f*, fin. *r*, rod or notochord, the first faint indication of a backbone. *nv*, nerve cord. *g*, gills. *h*, hole out of which water passes from the gills. *v*, vent for refuse of food.

small insignificant animal is called the "Lancelet,"[4] because it is shaped something like the head of a lance, and it is in many ways so imperfect that naturalists believe it to be a degraded form, like the acorn-barnacle; that is to say, that it has probably lost some of the parts which its ancestors once possessed. But in any case it is the most simple backboned animal we have, and shows us how the first feeble forms may have lived.

Flitting about in the water near the shore, eating the minute creatures which come in his way, this small fish-like animal is so colourless, and works his way down in the sand so fast at the slightest alarm, that few people ever see him, and when they do are far more likely to take him, as the naturalist Pallas did, for an imperfect snail than a vertebrate animal. He has no head, and it is only by his open mouth (*m*), surrounded by lashes with which he drives in the microscopic animals, that you can tell where his head ought to be. Two little spots (*e*) above his mouth are his feeble eyes, and one little pit (*n*) with a nerve running to it is all he has to smell with. He has no pairs of fins such as we find in most fishes, but only a delicate flap (*f*) on his back

4   Amphioxus lanceolatus (*amphi* both, *oxus* sharp).

and round his tail; neither has he any true breathing-gills, but he gulps in water at his mouth, and passes it through slits in his throat into a kind of chamber, and from there out at a hole (*h*) below. Lastly, he has no true heart, and it is only by the throbbing of the veins themselves that his colourless blood is sent along the bars between the slits, so that it takes up air out of the water as it passes.

But where is his backbone? Truly it is only by courtesy that we can call him a backboned animal, for all he has is a cord of gristle, *r r*, pointed at both ends, which stretches all along the middle of his body above his long narrow stomach, while above this again is another cord containing his nerve-telegraph (*nv.*) All other backboned animals that we know of have brains; but, as we have seen, he has no head, and his nerve-cord has only a slight bulge just before it comes to a point above his mouth. Now when the higher backboned animals are only just beginning to form out of the egg, their backbone (which afterwards becomes hard and jointed) is just like this gristly rod or *notochord* (*r r*) of the lancelet, with the spinal cord (*nv*) lying above it; so that this lowest backboned animal lives all his life in that simple state out of which the higher animals very soon grow.

This imperfect little lancelet has a great interest for us, because of his extremely simple structure and the slits in his throat through which he breathes. You will remember that when we spoke of the elastic-ringed animals in *Life and her Children*, we found that the free worms were very active sensitive creatures, whose bodies were made up of segments, each with a double pair of appendages; the whole being strung together, as it were, upon a feeding tube and a line of nerve-telegraph, but without any backbone. Now among these worms we find many curious varieties; some

have the nerve-lines at the sides instead of below, and one
sea-worm, instead of breathing by outside gills like the others, has slits in its throat through which the water can pass,
and so its blood is purified.

You may ask, What this has to do with backboned animals?
Nothing directly, but these odd worms are like fingerposts in
a deserted and grass-grown country, showing where roads
may once have been. The lancelet, like the worm, has a line of
nerve-telegraph and a feeding-tube, only with him the nerve-
telegraph lies above instead of below. He has also slits in his
throat for breathing, only they are covered by a pouch. Thus
he is so different from the worms that we cannot call them
relations; but at the same time he is in many ways so like,
that we ask ourselves whether his ancestors and those of the
worms may not have been relations.

But you will say he is quite different in having a gristly
cord. True—but we shall find that even this does not give us a
sharp line of division. By looking carefully upon the seaweed
and rocks just beyond low tide, we may often find some curious
small creatures upon them, called Sea-Squirts or Ascidians (B,
Fig. 2).[5] These creatures are shaped very like double-necked
bottles, and they stand fixed to the rock with their necks
stretching up into the water. Through one neck (m) they take
water in, and after filtering it through a kind of net so as to
catch the microscopic animals in it and taking the air out of it,
they send it out through the other neck, thus gaining the name
of sea-squirts. So far, they are certainly boneless animals. But
they were not always stationary, as you see them fixed to the
rock. In their babyhood they were tiny swimming creatures
with tails (A and a), and in the tail was a gristly cord (r), with

5   For this drawing, and also those of Figures 1 and 4, I am indebted to Professor
A. C. Haddon; the larval form A is the young of Clavelina, found at Torquay.

Fig. 2.

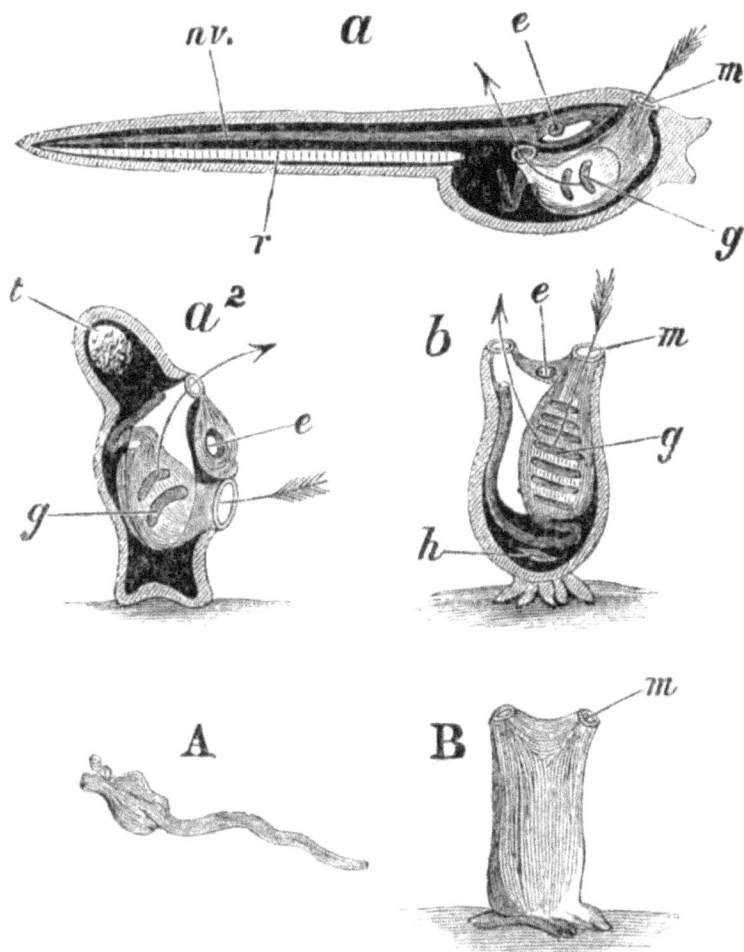

Diagram of the growth of a Sea-Squirt or Ascidian.
A *a*, Young free swimming stage. *a²*, Intermediate stage when first settling
down. B *b*, Full-grown Sea-Squirt.
m, mouth; *e*, hollow brain with eye; *g*, gill slits; *h*, heart; *r*, rod of gristle in
free swimming form; *nv*, nerve cord in same; t, tail in process of absorption
in intermediate form.

a nerve cord (*nv*) above it, like those we find in the lancelet. For this reason we were obliged to pass them by among the lower forms of life, because, having this cord (*r*), they did not truly belong to the animals without backbones; and yet now we can scarcely admit them here, because when they are grown up they are not backboned animals. They belong, in fact, to a kind of "No Man's Land," behaving in many ways like the lancelet when they are young, as if they had once tried to be backboned; and yet they fall back as they grow up into invertebrate animals.

So we begin to see that there may have been a time when backbones had not gained quite a firm footing, and our lancelet, with his friends the sea-squirts, seems to lie very near the threshold of backbone life.

And now that we are once started fairly on our road, let us turn aside before beginning the history of the great fish-world and pay a visit to a little creature whose name, at least, we all know well, and which stands half-way between the lancelet and the true fish. This is the Lamprey, represented by two kinds; the large Sea-Lamprey, caught by the fishermen for bait as it wanders up the rivers to lay its eggs, and the true River-Lamprey or Lampern, which rarely visits the sea.

What country boy is there who has not hunted in the mud of the rivers or streams for these bright-eyed eel-like fish, with no fins, and a fringe on back and tail? If you feel about for them in the mud they will often come up clinging to your hand with their round sucker-mouth, while the water trickles out of the seven little holes on each side of their heads. The small river-lampreys do not hurt in the least as they cling, though the inside of their mouth is filled with small horny teeth. But the larger sea-lamprey uses these teeth as sharp weapons, scraping off the flesh of fish for food as he clings to them.

Fig. 3.

Figure of a full-grown Lamprey[6] and of the young
Lamprey, formerly called Ammocœtes.

Showing the seven holes through which it takes in water to breathe.

These Lampreys, together with some strange creatures,
the "Hags" or "Borers,"[7] belong to quite a peculiar family,
called the Round-mouthed fishes,[8] and, though they stand
much higher in the world than the lancelet, yet they are
very different from true fish. Like the lancelet they have
only a gristly cord for a backbone, but this cord has begun
to form arches over the nerve battery, and it swells out at
the end into a gristly skull covering a true brain. They have

6   Petromyzon (*petra*, stone; *myzo*, to suck).
7   Myxine.
8   Cyclostomata (*cyclos*, circle; *stoma*, mouth).

clear bright eyes too, and ears, which if not very sharp, are at least such as they can hear with; they have only one nostril, and their mouth is both curious and useful. When it is shut it looks like a straight slit, but when it is open it forms a round sucker with a border of gristle, and this sucker clings firmly to anything against which it is pressed, so that a stone weighing twelve pounds has been lifted by taking a lamprey by the tail. Inside the mouth the palate and tongue are covered with small horny teeth, and these are the lamprey's weapons.

Salmon have been caught in the rivers with lampreys hanging to them, and where the mouth has been the salmon's flesh is rasped away, though he does not seem much to mind it.

Lastly, the lamprey has a peculiar way of breathing. He has seven little holes on each side of his head, reminding us of the slits in the worm's throat and those hidden under the skin of the lancelet, and behind these holes are seven little pouches lined with blood-vessels, which take up air out of the water. These pouches are all separate, but they open by one tube into his throat. When the lamprey is swimming about it is possible that he may gulp water in at his mouth and send it out at the slits. But when he is clinging to anything he certainly sends water both in and out at the slits, so that he can still breathe, though his mouth is otherwise occupied.

And now, what is the history of his life? For three years he lives as a stupid little creature, with a toothless mouth surrounded by feelers, and tiny eyes covered over with skin, and he is so unlike a lamprey that for a long time naturalists thought he was a different animal and called him *Ammocœtes*. But at the end of the three years he changes his shape, and then he is as bright and intelligent as he was dull and heavy before. His one thought is to find a mate and help her to cover up her eggs. To do this a number of lampreys find their way up the river and set to work. Sometimes one pair go alone, sometimes several together, and they twirl round and round so as to make a hole in the sand, lifting even heavy stones out with their mouths if they come in the way. Then they shed the spawn into the hole, where it is soon covered with sand and mud, to lie till it is safely hatched, and when this is done the marine lampreys swim out to sea to feed on the numberless small creatures in it, or to fasten upon some unfortunate fish.

But there are round-mouthed fishes even more greedy than these. It is not only among the lower forms of life that some creatures, such as worms, which are driven from the outer world, find a refuge inside other animals. But here again we meet with the same thing, for those relations of the lampreys, the hags or borers, which we mentioned above, use their sharp teeth to bore their way into other fish so as to feed upon them. These greedy little creatures actually drill holes in the flesh of the cod or haddock and other fish, and eat out the inside of their bodies, so that a haddock has been found with nothing but the skin and skeleton remaining while six fat hags lay comfortably inside.

So the round-mouthed fishes, feeble though they are, hold their own in the world. How long ago it is since they first began the battle of life we shall probably never know for certain; but if some little horny teeth[9] found in very ancient rocks belong to their ancestors, they were most likely among the first backboned animals on our globe.

9   Called *conodonts*, and found in Lower Silurian rocks earlier than any bones of true fish.

At any rate they are very interesting to us now, for they
have wandered far away from the true fishes, and give us
a glimpse of some of the strange by-paths which the backboned animals have followed in order to win for themselves
a place in the race for life.

THE ANCIENT FISH & THEIR HVGE RIVAL

CHAPTER II.

HOW THE QUAINT OLD FISHES OF ANCIENT TIMES HAVE LIVED ON INTO OUR DAY.

WHO is there among my readers who wishes to understand the pleasures, the difficulties, and the secrets of fish life? Whoever he may be he must not be content with merely looking down into the water, as one peeps into a looking-glass, or he may, perchance, only see there the reflection of his own thoughts and ideas, and learn very little of how the fishes really feel and live. No! if we want really to understand fish-life we must forget for a time that we are land and air-breathing animals, and must plunge in imagination into the cool river or the open sea, and wander about as if the water were our true home. For the fish know no more about our land-world than we do about their beauti-ful ocean-home. To them the water is the beginning and end of everything, and if they come to the top every now and then for a short air-bath they return very quickly for fear of being suffocated. Their great kingdom is the sea—the deep-sea, where strange phosphorescent fish live, lying in the dark mysterious valleys where even sharks and sword-fish rarely venture;—the open sea, where they roam over wide plains when the ocean-bottom makes a fine feeding-ground, or where they thread their way through forests of seaweed, while others swim nearer the surface and come up to bask in the sun or rest on a bank of floating weed;—and the shallow sea, where they come to lay their eggs and bring up their young ones, and out of which many of them venture up the mouths of rivers, while others have learnt to remain in them and make the fresh water their home.

The tender little minnows that bask in the sunny shallows of the river have never even seen the sea, their ancestors left it so long long ago; yet to them, too, water is life and breath and everything. The green meadow through which the river flows is just the border of their world and noth-ing more, and the air is boundless space, which they never visit except for a moment to snap at a tiny fly, or when they jump up to escape the jaws of some bigger fish. Every one knows the minnow, and we cannot do better than take him

as our type of a fish in order to understand how they live
and move and breathe. Go and lie down quietly some day by
the side of the clear pebbly shallows of some swiftly-flowing
river where these delicate little fish are to be seen; but keep
very still, for the slightest movement is instantly detected.
There they lie

> "Staying their wavy bodies 'gainst the streams
> To taste the luxury of sunny beams
> Tempered with coolness. How they ever wrestle
> With their own sweet delight, and ever nestle
> Their silver bellies on the pebbly sand!
> If you but scantily hold out the hand,
> That very instant not one will remain;
> But turn your eye and they are there again."

If you can be motionless and not frighten them you may
see a good deal, for while some are dashing to and fro, others,
with just a lazy wave of the tail and the tiny fins, will loiter
along the sides of the stream, where you may examine their
half-transparent bodies. Look first at one of the larger ones,
whose parts are easily seen, and notice how every moment he
gulps with his mouth, while at the same time a little scaly cover
($g c$, B, Fig. 4) on each side of the head, just behind the eye,
opens and closes, showing a red streak within. This is how he
breathes. He takes in water at his mouth, and instead of swallowing it passes it through some bony toothed slits (g, A Fig.
4) in his throat into a little chamber under that scaly cover; in
that chamber, fastened to the bony slits, are a number of folds
of flesh full of blood-vessels, which take up the air out of the
water; and when this is done he closes the toothed slits and so
forces the bad water out from under the scaly cover back into
the river again. It is the little heart (h), lying just behind the

Fig. 4.

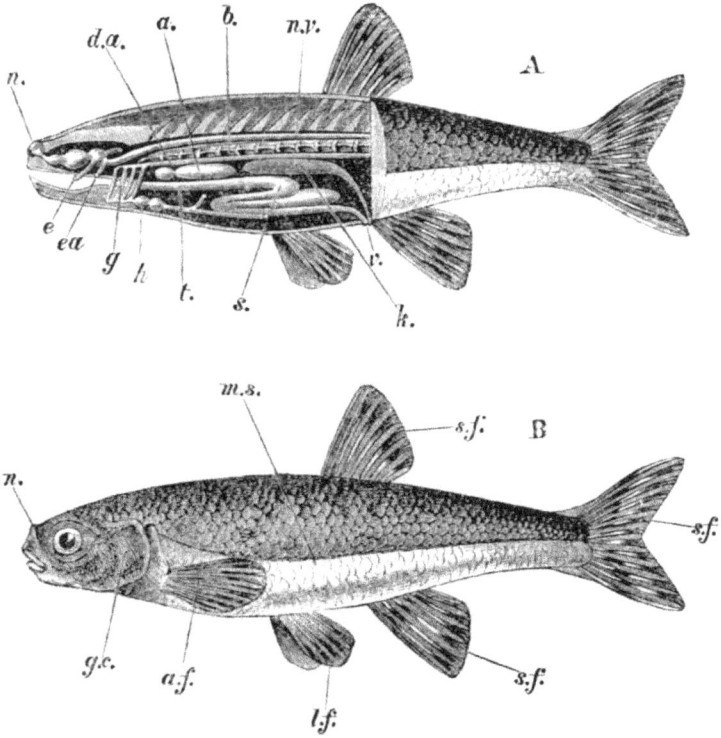

The structure of the Minnow and the living fish.

A *n*, nose-pit; *e*, eye-nerve; *ea*, ear-nerve; *g*, gills; *h*, heart; *t*, food-tube; *s*, stomach; *k*, kidney; *v*, vent; *da*, dorsal-artery; *a*, air-bladder; *b*, backbone; *nv*, nerve cord or spinal cord.

B *n*, nose; *gc*, gill cover; *af*, arm-fin; *lf*, leg-fins; *sf*, single fins; *ms*, mucous scales.

gills, which pumps the blood into the channels in those red folds, and as it keeps sending more and more, that which is freshened is forced on and flows through the rest of the body. It goes on its way slowly, because a fish's heart has only two chambers instead of four as we have, and these are both employed in pumping the blood *into* the gills, so that for the rest of the journey through the body it has no further help. For this reason, and also because taking up air out of the water is a slow matter, fish are *cold-blooded* animals, not much warmer than the water in which they are.

But while our minnow breathes he also swims. He is hardly still for a moment, even though he may give only the tiniest wave with his tail and fins, and he slips through the water with great ease, because his body is narrow and tapers more or less at both ends like a boat. At times, too, if he is frightened, he bounds with one lash of his tail right across the river; and if you look at one of the small transparent minnows you will see that he has power to do this because his real body, composed of his head and gills, heart and stomach, ends at half his length (see Fig. 4, A), and all the rest is tail, made of backbone and strong muscles, with which he can strike firmly. This is one great secret of fish strength, that nearly one half of their body is an implement for driving them through the water and guiding them on their way. Still although the tail is his chief propeller, our minnow could not keep his balance at all if it were not for his arm and leg fins. You will notice that it is the pair of front fins (*af*) which move most, while the under ones (*lf*) are pressed together and almost still. Besides these two pairs he has three single fins (*sf*), one under his body, one large V-shaped one at the end of his tail, and another single one upon his back. All these different fins help to guide him on his way; but while the single ones are fish-fringes, as it were, like the fringe round the lancelet's body, only split into several parts, the two pair under his body are real limbs, answering to the two pair of limbs we find in all backboned animals, whether they are all four fins, or all four legs, or wings and legs, or arms and legs.

These paired wings are most important to the minnow, for, if his arm-fins were cut off, his head would go down at once, or, if one of them was gone he would fall on one side, while, if he had lost his fins altogether, he would float upside down as a dead fish does, for his back is the heaviest part of his body. It is worth while to watch how cannily he uses them. If you cannot see him in the stream you can do so quite well in a little glass bowl, as I have him before me now. If he wants to go to the left he strikes to the right with his tail and moves his right arm-fin, closing down the left, or if he wants to go to the right he does just the opposite; though often it is enough to strike with his tail and single fin below, and then he uses both the front fins at once to press forward.

But how does he manage to float so quietly in the water, almost without moving his fins? If your minnow is young and transparent you will be able to answer this question by looking at his body just under his backbone, and between it and his stomach. There you will see a long, narrow, silvery tube (a, Fig. 4) drawn together in the middle so that the front half near his eyes looks like a large globule of quicksilver, and the hinder half like a tiny silver sausage. This silvery tube is a bladder full of gas, chiefly nitrogen, and is called the *air-bladder*. Its use has long been a great puzzle to naturalists, and even now there is much to be learnt about it. But one thing is certain, and that is, that fish such as sharks, rays, and soles, which have no air-bladders, are always heavier than the water, and must make a swimming effort to prevent sinking. Fish, on the contrary, which have air-bladders, can always find some one depth in the water at which they can remain without falling or rising, and we shall see later on that this has a great deal to do with the different depths at which certain fish live. Our minnow floats naturally not far from the top, and, even if he were forced to live farther down, the gas in his bladder would accommodate itself after a few hours if the change was not too great, and he would float comfortably again.

And now the question remains, What intelligence has the minnow to guide him in all these movements? If you will keep minnows and feed them yourself every day you will soon find out that they see, smell, and feel very quickly, though their hearing and taste are not so acute. They are cunning enough too, and will often steal a march upon heavier and slower fish, snatching delicate morsels from under their very noses. For our little minnow can boast of a real brain, though it is a small one in comparison with his size. All along, above his delicate backbone, the thread of nerve telegraph (*nv*, Fig. 4) runs under protecting bony arches, and sends out nerves on all sides to the body and fins; and when it reaches the head it swells out, under a bony covering, into a small brain, sending out two nerves to the ears (*ea*), in front of which is a second part, with two nerve-stations (*e*) for the eyes, and beyond this a third part, with two more for the nostrils, besides others which go to the face. Look on the top of a minnow's head and you will see two little raised bumps (*n*). These are its nostrils, but remember *they have nothing to do with breathing*; they have not even any connection with the mouth, but are simply little covered cups, each with two openings for water to flow in and out, and they are lined with nerves, which, tickled by good or bad scents in the water, carry to the brain a warning, or a promise of good things.

Such, then, is our little minnow, and the different parts of his body are supported by a slender bony-jointed backbone, with ribs growing from it, supporting a strong mass of flesh on his sides. He is a delicate tender creature, but is protected and buoyed up by the water, out of which he never attempts to go. The thin, rounded, transparent scales which cover his body, growing out of little pockets in his skin, just like our nails on the tips of our fingers, protect this skin from the water and from rough treatment; while they themselves are kept soft by a slimy fluid which oozes out from under them, and especially through the dark line of larger scales (*ms.* Fig. 4) running along his body.

* * * * *

Now the minnow is a bony fish, and from it we can learn very fairly what bony or modern fishes are like. But these fish were not the founders of the race; long before they existed there was another very ancient group of fishes in the world, which were in many ways more like the lancelet and the lamprey; and to find such descendants of this ancient group as are now living we must leave the river and find our way into the open sea.

If we do this, we shall travel not many miles from the shore in summer, wending our way through shrimps and lobsters, gurnards, cod-fish, soles, and turbot, before we may chance to come across a great Blue shark, with his slaty-coloured back and fins, swimming heavily but strongly through the water, and turning sharply from time to time to seize a passing fish, his white belly gleaming like a flash of light as it comes uppermost, and then disappearing again in the dark water.

> "His jaws horrific, armed with threefold fate,
> Here dwells the direful shark."

Or if this formidable monster does not happen to be in the neighbourhood, another kind, the Dogfish, may cross our path, perhaps the Smooth hound, crushing the crabs and lobsters in his tooth-lined mouth, or the Rough hound fastening her purse-like egg to the seaweed by its long string-like tendrils; or, farther out still, we may perhaps see the Thresher shark lashing the water with his long pointed tail, to drive the frightened fish within his reach; or, if we were off the west coast of Ireland, the huge but harmless Basking shark might be floating calmly by in the warm sunshine. For sharks travel all over the ocean, and though they prefer the warm seas, where they sometimes reach a size of forty feet long and more, yet many of the smaller kinds visit our coasts in summer.

Fig. 5.

The Blue Shark[10] (*FROM BREHM*).

To show the five slits in the neck, the uneven tail, and the mouth opening under the pointed snout.

10   Carcharias glaucus.

Now, at first sight we might imagine that these huge
monsters, the terrible tyrants of the sea, must be the last
and most finished production of fish-life; but if we look a little
closer we shall be undeceived. Examine a shark in any good
museum, and you cannot fail to be struck with his strange
form. Look first at his tub-like body, so different from the narrow wedge-shape of the minnow, the herring, or the salmon.
Then observe his skin, which is either tough, more like that
of other animals, or thickly covered with short blunt teeth,
which sometimes, especially in front of the fins, become long
pointed spines. There is no trace of fish-scales here. Look at
his mouth opening under his pointed snout, and you will see
that as the skin turns over the lips these blunt teeth line his
mouth, so that he has several rows fit for biting, and they are
sometimes so formidable that they can cut a man in two at
one snap. Then look more especially at the sides of his throat,
and there you will see on each side from five to seven slits,
reminding you at once of the slits of the lamprey, though they
are long instead of round. For the shark has never arrived at
having true gills under a horny cover like the minnow, but still
breathes by pouches and slits somewhat after the way of the
lowly round-mouthed fishes. Lastly, observe his curious tail.
In nearly all living fish the tail is even[11] or V-shaped, but in
the sharks the top point is usually longer than the lower one,[12]
and in some, such as the Thresher, it is very remarkably so.

This uneven tail is the badge of a very ancient race; out
of the shark family we scarcely find it anywhere now except
among the sturgeons, who, we shall see, are old-fashioned too.

And now when we inquire into the growth of the shark and
the kind of backbone he has, we find that he has still more links

11   Homocercal.
12   Heterocercal.

with the lower fish-like animals. For when he is young he has nothing but a rod of gristle or cartilage running between the long narrow feeding-tube and the spinal cord; but this rod is flattened in front, and as the young shark grows up the flat part enlarges so as to form a boat-like box—the skull, round the swollen end of the nerve telegraph—the brain. Meanwhile the rod becomes divided into rings, and from each ring an arch of gristle growing upwards surrounds the nerve cord so as to protect it from injury, and the whole skeleton becomes firm and strong. But though the shark is one of the strongest of sea-animals *he never loses this gristly state of his backbone or his skeleton*; however much he may strengthen it by hard matter it never becomes true bone, but remains quite distinct from the skeleton of the bony or osseous fish.

Fig. 6. - The Sturgeon[1] entering a Russian river.

---

1   Acipenser sturio.

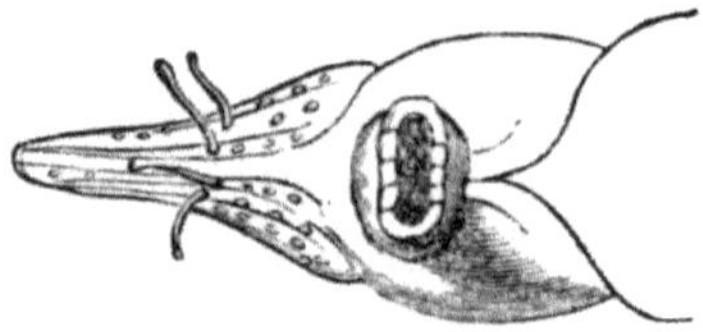

Fig 7. - The Sturgeon's head seen from below, showing the tube-like mouth and the four barbels or feelers.

We see, then, that there is a race of gristly or *cartilaginous* fishes, which, though they have grown strong and powerful, still hold to many primitive habits in forming both their body and skeleton. Nor do the sharks stand alone, for the large sturgeons, which live partly in the sea and partly in fresh water, crowding up the rivers of Russia and America to grope in the river mud for food, and to lay their millions of eggs, are also remnants of the ancient type. It is true that with them the slits in the neck are covered by a horny flap like the bony fish, and like them too they have an air-bladder under the backbone.[13] But they too have a gristly skeleton, and the gristly rod more or less hardened runs right along their back. In other respects they are perhaps even more peculiar than the sharks; for the sturgeon's head is covered with hard bony shields, and five rows of bony bucklers are arranged along his body. We seem almost to have got back among the armour-covered animals as we look at his shiny plates, reminding us that with a mere gristly skeleton within, it may have been wise for the early types of fish to wear some outward protection. His snout is long and pointed, with four delicate feelers hanging down from it, and his mouth, which is quite under his head, is a soft open tube without teeth, which he can draw up or push out to suck up fish or any animal matter he finds in the mud.

Clearly the sturgeon is an old-fashioned fellow, as you may see for yourself, when specimens caught at the mouths of our rivers are shown in the fishmongers' shops. I have often

13   Isinglass is made from the covering of this air-bladder.

wondered, when standing looking at him and at the sharks in the British Museum, whether the people who stroll by have any idea what a strange history these quaint old fishes have, or how they stand there among the scaly and bony fishes lying in the cases around, just as an Egyptian and a Chinaman might stand in an English crowd, descendants of old and noble races of long long ago, whose first ancestors have been lost in the dim darkness of ages, whose day of strength and glory was at a time when the modern races had not yet begun to be, and whose representatives now live in a world which has almost forgotten them.

In the silent depths of the large lakes of North America there is a fish called the Bony pike,[14] a huge fellow often six feet long, with a long beak-shaped mouth, which he snaps as he goes, devouring everything that comes in his way. This fish has his body covered with lozenge-shaped, bony, enamelled scales, like the fish of long ago, and so too has the strange Bichir,[15] which wanders above the cataracts of the Nile, with its row of eight to eighteen fins raised upon its back like tiny sails. Then again there are the curious calf-fish of North America,[16] of the Amazons,[17] of the Nile,[18] and of the rivers of Queensland in Australia.[19] These all have gristly skeletons, and together with the sharks and sturgeons make up all that remains of those strange shadows of the past moving among the bony fishes of to-day.

The mud-fishes are indeed the most curious of all, for they breathe both water and air, and in the Nile and Gambia often coil themselves round in the mud when the water goes

14   Lepidosteus.
15   Polypterus.
16   Amia.
17   Lepidosiren.
18   Protopterus.
19   Ceratodus.

Fig. 8.

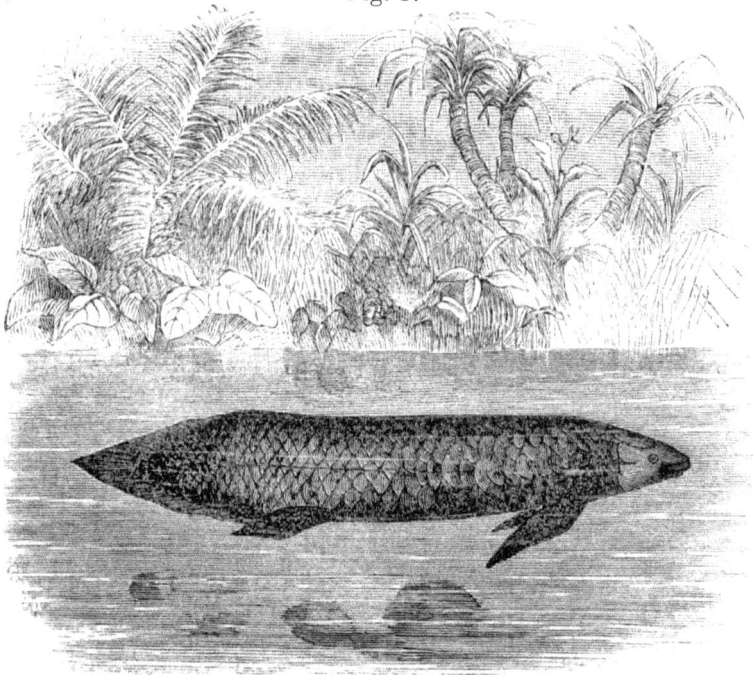

The Ceratodus of Queensland, an air-breathing and water-
breathing mud-fish of the ancient type, with paddle-fins.

down, and, lining their bed with slime, sleep comfortably till
the rains refill the pools with water.[20] The fact is they have
two quite separate ways of breathing. They have gills with
which they can take air out of the water like other fish, and
these they always use when they can. But they have a tube
in their throat leading into the air-bladder lying under their
backbone, and through this they can breathe in air when
they cannot get it from the water. In *Amia* especially, which
is a true enamel-scaled fish, this air-bladder is divided into
numerous cells, and it breathes with it just as with a lung.

It was in the year 1870 that the *Ceratodus*, or the "Bar-
ramunda," as the Australian natives call him, was discovered

20   These fish, coiled round, may be seen in the British Museum.

in the rivers of Queensland; and since then he has become very famous, for, more than any of the others, he is like the fishes of long ago. He is a lumpy fish, sometimes as much as six feet long, with a gristly cord for a backbone. His body is covered with large rounded scales, and he has a broad fringe round his pointed tail. His fins are more like paddles than fish fins, having several joints, and he uses them, together with his fringed tail, to flap along in the water, or even to wander over the reedy flats at night, chewing the weeds with his broad ridged teeth. And as he flaps along, from time to time, when the water is too muddy for his liking, he comes up to the top, and with a great gulp swallows air into the air-chamber. But before he can do this he must send out the bad air within, and in doing so he gives a grunt which is often heard far away at night in those still Australian wilds. He need not come up for air-breathing, however, if the water is pure, for, strange to say, the whole course of his blood can be altered to suit his wants. When he can get clear water to breathe through his gills the blood flows to them to be freshened, and his air-bladder simply takes in gas from the body as it does in other fishes, and wants feeding with good blood. But when he comes up to breathe then the blood is carried the other way, and comes to the air-bladder to be freshened.

And now if we want to read the history of all these strange forms, you must let me take you by the hand and lead you in imagination back, back through millions of years, to a time so long ago that we cannot even count the ages between. As we recede from our own day we shall leave behind us all the kinds of plants and animals we now know so well, and meet with strange kinds only bearing a general resemblance to them. After a long journey of thousands and thousands of years, in which the plants and animals, and even the very shape of the continents and islands, have gone through many changes, we shall get back to the time when the lime-builders were forming thin layers of chalk at the bottom of the sea, which were afterwards to become our enormous chalk hills. Still backwards we must go through all that long period, and then through three others quite as long, with ever-changing scenes of life and climate and geography, till we find ourselves in those grand old forests whose trees and plants we now dig out as coal.

Even then we must not stop to rest, though we are getting back to the dim ages of the world, for the journey is not yet ended. On, on, backwards through countless years, till we lose sight not only of beasts and birds and reptiles, but even of insects and flowering plants, which, at the time we are reaching, had not yet begun to be. At last we lose almost all life upon the land, so far as we can tell, and after another long period has passed before us we find ourselves in a scene of *water, water everywhere*.

True, there is a line of shore where strange ferns and unknown club-mosses and reed-like plants are growing; but these only border the vast water-world, and we have reason to believe that no living animal wanders over that wild and barren country. But the water itself is full of life, though its inhabitants are of low kinds, as if Nature herself was as yet only half-awake.

Rich and rare seaweeds carpet the floor of the ocean, mingled with delicate flint-sponges and old massive corals; beautiful feather-stars in the form of rooted stone-lilies wave their slender arms; greedy star-fish, grazing sea-urchins, and all their many relations, grope upon the rocks; and sea-snails crawl or float in countless numbers. The Nautilus, too, is there, with curious half-uncoiled companions of forms we have never seen before; and huge sea-woodlice, the Trilobites of olden time with their three-ridged shields, burrow in the sand, or roll themselves up at the bottom of the water. And above all these, among many kinds of armour-covered animals, a huge form, nine feet long, like a lobster, with an imperfect head, rows himself along with his oar-like hind feet, seizing the smaller creatures with his long nipping claws in front. For we have travelled back to a time when the crustaceans were the most powerful animals in the world, and the huge lobster-like Pterygotus was the monarch of the seas.

It was in the midst of a scene such as this that we first find the feeble ancestors of the Sturgeon and the Shark beginning to make their way in the world. It may be that creatures such as the sea-squirts, the lancelet, and the lamprey, were there to bear them company, but these soft animals could leave no trace behind except the tiny teeth of the lampreys; for they had no enamelled plates like the plated fish, no hard teeth-spines like the sharks, which could become buried in the soft mud when they died, and remain, together with the hard shell of their enemy the Pterygotus, to be dug out now in our day and bear witness to the fight they fought. But the plated-scaled fish had something to leave behind, and from their remains we can picture to ourselves a group of clumsy fish scarcely a foot long, with hind fins like paddles and single-fringe fins on their back, with enamelled lozenge-shaped plates on their bodies and unevenly pointed tails. These fish would keep well out of the way of the Pterygotus, because they were small and weak and he was large and strong. We may imagine them gliding among the seaweeds, and hugging the shore as they chewed the plants with their flat-ridged teeth, for their skeletons were probably feeble and their armour-like shields were heavy, and they would not be so active as the little shark-like animals, not bigger than a half-pound perch, with tough skins and sharp spines, which swam more boldly out to sea. These more active fish were the founders of the shark group, and those sharp spines, together sometimes with the tough skin, remained buried in the mud, and have come down to us as fossils.[21]

We should find it difficult to say exactly to what class all these early fish belonged, for there were very few kinds, and therefore fewer distinctions, between them in those days; and many peculiarities which afterwards appear in different groups either did not exist or were united in one fish. It is enough for us that they were the ancestors of our sharks and sturgeons and mud-fish of to-day; and though they were but small and weak, yet they were the beginning of a powerful race of creatures, for they had the great advantage of a growing inside skeleton, which could vary and strengthen with their bodies from generation to generation, while their rivals, the Pterygotus and his companions, had only their heavy cumbrous armour with a mass of soft flesh inside, and were but lumbering creatures at best.

And so we find that as thousands and thousands of years rolled by, the descendants of the enamel-shielded fish began to improve, and became larger and more powerful as the generations passed on, till they became masters of the shallow seas, and after awhile of the rivers and lakes. By the time that the first air-breathing creatures, the May-flies and Dragon-flies, had found their way out of the water into the forests of pines and tree-ferns on the land, and left their tender wings in the soft ground of the ponds and lakes, large

21   Ichthyodorulites.

fishes[22] whose tails were uneven-pointed like the sturgeon,—whose bodies were covered with lozenge-shaped enamelled scales and their heads with shields,—were grazing along the shores and in the rivers and bays, with probably swarms of smaller kinds which have left no traces behind.

These were peaceable fish which fed upon plants, and among them were some curious forms with paddle-like fins and broad-ridged teeth, which, as they swam under the shade of the huge forest trees, would come to the top and take in air through their mouth. These were the distant ancestors of our present mud-fishes, and through all the passing ages, from the time of the coal forests till now, they have kept their fish-like form, so that we have their descendants among us now in the Australian *Ceratodus* and the mud-loving *Protopterus* of the Nile.

But besides these gentle vegetarians there were in the sea huge enamel-scaled monsters, with terrible jaws and gigantic teeth, floundering about and making great havoc among the crab-like animals. One of these, whose head-shield has been found in the ancient rocks of Ohio in America,[23] must have been at least fifteen feet long, with a huge head, three feet long and a foot and a half broad; and no doubt there were many others like him, having a fine time of it now that they were the strongest creatures living. For this was the Golden Age of fishes, just before the time when the coal-forests grew; and the clumsy crab-like animals, and the trilobites, which had had their innings when the fish were small, now began gradually to be exterminated by their powerful enemies. Little by little they gave up the battle of

22   Dipterus.
23   Dinichthys.

life, and the larger ones died out altogether, leaving only those smaller crustaceans which did not clash with the fish.

So time passed on. The coal-forests grew, and died away and were buried; and as the ages rolled by a still stronger class of animals began to grow up which was to pay back upon the enamel-scaled fish the vengeance they had wreaked upon the crustaceans. For in the coal forests we first meet with creatures like our newts and salamanders, and after these came the true air-breathing reptiles (see Chap. v.), which swarmed over land and sea. There were the fish-lizards, with their strong swimming paddles and sharp teeth, and the swan-like lizards, with their long necks, which enabled them to strike their prey in the water; and these, together with the flying-lizards, and the huge dragon-like reptiles which haunted the shore, made the life of the heavily-moving enamelled fish a burden to them. So they, in their turn, began to give way, and became smaller and rarer as the history went on, till at the time when the chalk-building animals were at work at the bottom of the sea we begin to lose sight of all but those few forms which linger still. It was about this time that the Sturgeon, as we now know him, became the chief representative of these old cartilaginous fishes, and to this day he and his children go on travelling up the rivers of Europe, Asia, and America, or crossing from sea to sea—a living example of those ancient races which ruled the seas of long ago.

The history of the small shark-like animals was rather different. They too grew strong and powerful before the reptiles came, and they did not afterwards lose much of their greatness. With the wide ocean for their home, and not troubled with the heavy enamelled plates of their companions, they kept clear of the monster reptiles, or struggled with them bravely. Some took to the open sea, and from them are descended the giant sharks of to-day which still remain masters of the ocean. Others still lingered near the shore, where we find quite new forms springing up; some, like the Chimæra or "King of the Herrings," formed a group of their own, half-way between sharks and sturgeons; and some, slightly flattened like the huge Monk-fish, hide themselves in the loose sand when seeking their prey. Others, the Skates and Rays, with flat bodies, and long tails serving as rudders, shoved smoothly along with a wavy flapping motion of their broad arm-fins. These too lie chiefly at the bottom of the sea, where their dusky colour hides them both from the fish they would wish to attack and those that would attack them; for while the sharks trust to their strength, the skates and rays trust to stratagem, and, coming along stealthily in the shadow, flap rapidly over their prey and suck them into their open mouth below. And for further protection we find some of them, such as the Sting-rays, armed with barbed spines; others, such as the Torpedo-fish, with electric batteries in their heads, which they can use to stun and kill their enemies; while others again, such as the Saw-fishes of the Tropics, have the front part of their skull lengthened out in a long bony weapon, armed with teeth, which they use to rip open the bodies of their prey.

All these formidable fish are descendants of the shark family, which, with powerful gristly backbones, strong fins and tails, and highly developed brains, refused to be suppressed as their plated companions were, but found room in the wide ocean to do battle for themselves, and improve in many ways upon their ancestors. They do not, like the sturgeon and the bony fish, lay their thousands of eggs, but are content with one or two at a time, such as the leathery purse-eggs of the skate and the rough hound shark; or give birth to a dozen or twenty living young ones. Yet they are so well fitted for their life that they flourish and keep their ground, so that while the enamel-scaled fish and the mud-fish are small groups, many of them fading away, the sharks and rays bid fair to be the race which will keep up the traditions of those quaint old Fishes of ancient times, which were once the masters of the world.

THE BONY FISH IN THE EARLY DAYS

CHAPTER III.

THE BONY FISH, AND HOW THEY HAVE SPREAD OVER SEA AND LAKE AND RIVER

WHEN the palmy days of the enamel-scaled fish had passed away, and the sharks and rays had taken up their various quarters in different parts of the sea, there still remained vast tracts and many snug nooks and bays admirably fitted for fish-life. But these were not empty, for long before this time another order of fish—light, strong, and active,—had been pressing forward to take possession of every vacant space.

If we could dive under the water and watch the fishes at home we should see at once how much more agile and easy the bony fish are in their movements than their gristly companions. Look at a shoal of silvery herrings as they swim and leap and gambol, or a fine salmon sailing up the river or springing over a waterfall, or a tiny stickleback darting across the stream, and compare their graceful motion with the ponderous though powerful movements of an unwieldy shark. Any one who has done this will feel at once that though the sharks have still kept their power as tyrants of the sea, because they are so strong and big, yet these light skirmishers are much more at their ease, and move with much less effort in the water, so that it is natural they should have made their way into all parts of the rivers and seas. But where have they come from? We know very little of their early history, but what little we do know leads us to think that long ago they branched off from the enamel-scaled fish, and struck out a path of their own to make the most of the watery world.

Turn back for a moment to our little minnow, and recall his tender backbone made of joints hollowed out before and behind, with cushions of gristle between; those cushions, when the minnow was growing out of the minnow egg, were one long gristly cord, like the cord of the sturgeon, and it was only as the minnow grew that the bony joints hardened round it and separated it. Moreover, that huge bony pike which we find now wandering in the American lakes has bony joints hollowed out like the minnow's, although by his enamel-scales and uneven tail we know him to be one of the ancient fishes. Some time or other, then, the sturgeon, the bony pike, and the modern minnow, must have had a common ancestor, though we should have to reach him through millions of generations. In the same way, too, we find the red-folded gills covered by a scaly lid, both in the sturgeon and the minnow, though in other ways they are not exactly alike; while even the V-shaped tail of the modern fish is not so different from the ancient shape as it seems, for the end of the backbone runs up into the top branch of the fork as it does in the uneven tails of the olden fish. Lastly, the delicate rounded scales on our minnow's body are not entirely the property of bony fishes, for we find such scales on the mud-fishes, the *Amia* and *Ceratodus* (see p. 36); while the little modern stickleback, on the other hand, has bony plates, reminding us of those of olden times. We see, then, that the bony fish still carry upon them many signs of their origin from the older fish, and when once the coast was left clear, and they got a fair start, we can easily imagine that the fish of this younger race which was still in its childhood, and easily moulded to suit different kinds of life, would press forward in every direction and make the most of every chance.

And so we find that little by little, from the time of those chalk seas till now, the remains of enamel-scaled fish grow rarer and rarer in the hardened mud, and the bones and scales of modern fish take their place, till this bony race has spread so far and wide that in our own day, if we were to start from the head of a river and swim down into the open sea of the Atlantic or Pacific, we should meet on our way bony fish of all shapes and sizes and habits of life. River-fish and lake-fish and sea-fish; shore-fish, surface-swimming fish, and fish of the deep sea; flat-fish like the sole, half hidden in the sand, and long rounded fish like the eel, threading their

way through holes and passages all over the world; flying
fish with long arm-fins, and clinging fish whose fins form a
sucking disk; nay, even so strange a thing as an angling-fish,
whose back fin is turned into a fishing-rod to attract his prey.

All these, during the long ages since they first started in
life, have been learning to make use of some area in the wide
expanse of water spread over our globe, and it remains for us
now to see how they have succeeded. Where shall we make
our start? If we begin at home in the rivers we should have
to work, as it were, backwards, for the sea is the chief home
of fishes, and the rivers only the refuge of a few stray kinds.
The sea-shore would be, perhaps, our truest starting-point,
but then we should have to travel two different ways. Will
it not be best to dive down first into the silent depths of the
ocean, and learn what little is known of those which have
taken refuge there? Thence we can rise up to the open sea,
from there swim in to the shore, and then up the rivers and
back to our own land-home.

It makes but little difference where we take our plunge
into the deep sea, for changes of climate are scarcely or not
at all known there, and the fish seem to wander over every
part. Wherever it may be, then,—let us say in the seas of
the Tropics, which have given us most of our specimens—
let us dive down, down, till we reach about 1800 feet (300
fathoms).

> ... "For who can know
> What creatures swim in secret depths below,
> Unnumber'd shoals glide thro' the cold abyss
> Unseen, and wanton in unenvied bliss."

We shall be groping more and more in darkness as we
go, for the sunlight scarcely reaches beyond 1000 feet,
and we have left its last rays behind us, and the water is

growing icy cold. How strange, then, that the first fish we meet should have large wide-open eyes! This is the Beryx,[24] shaped something like a perch, but about a foot and a half long, and genealogists ought to look at him with respect, for his ancestors (see heading of Chapter) are almost the oldest known bony fish, and lived in the chalk seas.

Has he come down here because the upper world was too rough for him? If so, he has found comparative stillness, for he is far beneath the turmoil of the waves, and only the slowly creeping currents make any movement around him. But he has not escaped from the struggle for life, for not only is a good-sized shark coming his way, but a huge monster of the bony race, six feet long,[25] with wide-opened jaw, sharp pointed teeth, and large keen eyes, is wandering near in search of prey, devouring large and small fish with great impartiality.

Still in the dense darkness the Beryx must surely escape? No! for, strangely enough, lights are travelling about in this midnight region. The monster himself carries lamps upon his body, and a shoal of small oblong fish, something of the size and shape of a gudgeon, come swimming by, carrying on their sides whole rows of shining spots giving out phosphorescent light; while not far off another fish, called in India the Bombay duck, glows all over, as if his whole body had been rubbed in phosphorus. Nay! so far as we know the Beryx himself is probably gleaming with light, for his body is covered with a large quantity of the same slimy fluid which makes the "Bombay duck" phosphorescent when he is freshly drawn out of the sea.

So these curious fish, living in eternal darkness except

24   See Frontispiece. 1, Chauliodus; 2, Harpodon or Bombay Duck; 3, Plagiodus; 4, Chiasmodus, with a Scopelus in its stomach; 6, Beryx; 8, Scopelus.

when they make an expedition to the surface, carry many of them their own lights; and as we go deeper still more and more of them are found with shining mother-of-pearl-like spots on their head, or sides, or tail, so that the very darkness is alive with light. What slaughter and hunting there is among them! for they all eat each other, and even their own young, there being no plants for any of them to feed on. There are the deep-sea cod-fish; strange forms with large heads, long tapering tails, and thread-like fins, chasing the smaller fish, and falling victims themselves to the fierce *Stomias* which comes sailing along with its row of glowing lights, and its huge sharp teeth, ready to seize its prey. Both these fish go down as deep as ten thousand feet and more, accompanied by another fish quite as ferocious, though only a foot long, with large teeth sticking out of its mouth like the tusks of a boar, and curious round spots, with lenses in them, on its side, which may be eyes, or may be lanterns to light it on its road; and among these luminous fishes are wriggling along the deep-sea Conger eels, with toothless mouths and elastic stomachs, swallowing large fish whole; while another curious cod-like fish, whose stomach can stretch to more than four times its natural size, draws itself over its prey just as a snake does, and carries it in the hanging bag till it is digested. And deeper yet in the dead calm water roam many fishes with delicate feelers hanging from their mouths, while their fins are slender and tapering, so that they feel their way along the still depths. Among these are the Ribbon-fish, twenty feet long but only a foot deep, and never more than two inches thick in any part, with their long rosy fins floating like ribbons back from their heads and from under the body.[25]

25   In drawing up this sketch of the deep sea I am almost entirely indebted to Dr.

Strange monsters are all these deep-sea fish, some of them living as much as 16,000 feet under the surface of the sea, so that if Switzerland were turned upside down in mid-ocean, the peak of Mont Blanc would not reach down to where they swim. Yet they are only modified forms of ordinary fish from the world above, which have become fitted to live under that vast pressure of water. Their skeletons, though bony and well-knit together at that depth, are fibrous and slight compared to those of their surface relations, for although they have to resist a weight of from two to sixteen tons pressing all round them, a ton weight being added for every thousand feet, no special strength is required, because the dense water permeates their whole structure, and the pressures are everywhere equal. It is the same with them as with the most delicate and fragile insects living in our atmosphere, the pressure of which would tear them to pieces if unbalanced by equal pressures within and without.

But when these deep-sea fishes are brought up quickly to the surface, the outside pressure no longer balances that inside, and so their tissues loosen and their whole framework starts apart, so that they almost fall to pieces at a touch; and their air-bladder, if they have one, expands so much as to force the stomach out of the mouth, turning them almost inside out. Neither are their lanterns a special creation for their use, but merely adaptations of that slimy fluid which we saw oozing from the scales of the minnow. In some of the deep-sea fish even the outer bones are filled with this fluid, and the line of scales along the side has large openings, so that the body is bathed in glowing slime. In others it collects in glands on the sides, making the phosphorescent spots.

Günther's masterly sketch of the deep-sea fish in his excellent work.

In this way the deep-sea fish have become fitted to make a home in the very heart of the ocean. Some with large eyes, seeing by means of their own and their neighbours' light, others with small eyes and delicate feelers, testing each step as they go, and feeding, probably, on the shower of minute sea-animals that falls continually from above; while some, like the Beryx, the Bombay Duck, and the light-carrying Scopelus, which live nearer the top, come up on still nights to feed at the surface of the sea.

Fig. 9.

Remoras[26] clinging by their sucking-disk to the under part of a shark.—(*ADAPTED FROM BREHM.*)

26   Echeneis remora.

And now, as we rise again from the dark still depths up to warm layers of the tropical seas into which the sun is pouring his penetrating rays, it may happen that a large dark body moves between us and the surface, as the Great Blue Shark, or one of his smaller relations, ploughs his way through the water. But what are these little dark brown fish, with round gaping mouths, which are hanging by the top of their head and back from under the shark's belly? (see Fig. 9). Where he goes they go with him, and, as they are borne along, they feed upon the tiny sea-animals among which they are carried so easily. These cunning passengers, of whose very existence the shark seems unconscious, are the Remoras, or sucking-fish. You would scarcely think that they belong by descent to the mackerel tribe, a strong-swimming, active, and almost warm-blooded group of fish, with a large supply of nerves and blood-vessels to their muscles, so that they swim boldly out to sea, and make more use of the open ocean than almost any other group. But among all tribes there will be some weak members, and these must live by stratagem. The little remora is a feeble swimmer, and, having to live out at sea, has acquired a curious sucker by which he clings to sharks, and whales, and even ships, so that he is carried along without exertion. Yet this sucker, again, is only a special adaptation of the back-fin, which, instead of being single, as in other mackerel, has its spines divided and bent, one set to the left, the other to the right, and joined by a double set of plates, surrounded by a fringe of skin. This forms an oval disk, and, as the remora glides along under the shark's belly, he presses the damp membrane against the fish, and, drawing together the muscles of the plates, clings as firmly as a limpet to a rock.

Nor is the remora the only companion of the shark—

"Bold in the front the little Pilot glides,
Averts each danger, every movement guides;"

for the little steel-blue striped Pilot-fish,[27] another distant
connection of the mackerel tribe,[28] is hovering around, feeding
upon the scraps of the shark's food, and finding protection in
his neighbourhood, though in olden times he was supposed
to protect the shark. A brave little fish this, which has succeeded in making the shark his friend: while near him he is
safe from other fishes.

And now, as we continue our way in the open sea, it is
nearly always forms more or less related to the mackerel tribe
which cross our path. The slender Bonito[29] and the heavier
Tunny[30] sometimes ten feet long, are hunting below or on
the surface, and the beautiful Dorados,[31] or gold-mackerel,
as the Germans call them, with their silvery blue backs
tinged with a sheen of gold, their dull-coloured fins, and
their golden eyes, are driving by in large shoals in pursuit
of the flying-fish. All these are powerful swimmers, and they
have no air-bladder, which is an advantage to such active
hunters which wish to turn rapidly, to go down deep or rise
to the top, and change their position at every moment; for in
all these movements a natural float inside is a hindrance to
be overcome. And so we find that in fish, even of the same
family, some have lost the air-bladder, while others have it
enlarged to meet their wants, as in the case of the lovely
blue and silver sun-fish[32] for example, which, though quite

27   Naucrates.
28   In this description I am not alluding simply to the mackerel *family* Scombridæ,
but to that much larger group *Cotto-Scombriformes*, to which so many ocean fish belong,
and even the sword-fish is allied.
29   Thynnus pelamys.
30   Thynnus thynnus.
31   Coryphæna.
32   Lampris luna.

near relations of the dorado, have very large air-bladders, enabling them to float quietly on the top of the water, waving their deep scarlet fins.

Fig. 10.

Flying-Fish[33] pursued by the Dorado.[34]

But while we are watching all these large and strong swimmers an active and bloodthirsty struggle is going on, for the bonitos and the dorados are looking to make their meal upon the little Flying-fish, which are straining every nerve to escape them, while here and there one drops down

33   Exocœtus.
34   Coryphæna.

into their very mouths. Lovely little creatures these are, of
the Pike family, which have taken to the open sea, where
they rise with a stroke of the tail many feet out of the water,
their bright purple backs and silvery sides gleaming in the
sun, as, with their long transparent arm-fins outspread,
they float for as much as two hundred yards before they
fall back, to spring up again with another stroke. Their air-
bladder, which is half as long as their body, and contains in
a six-inch fish as much as three and a half cubic inches of
gas, stands them in good stead, and they rise and fall with
quick rapid flights out of the reach of their foe, so that in the
open sea they do fairly well on the whole, though, if they
venture near land, the sea-birds persecute them in the air.
Nor do they stand alone in this curious habit of flying, or
rather floating, in the air, for a larger fish of quite another
family, the "Flying Gurnards,"[35] with a smaller but still
ample air-bladder, and long arm-fins, may also be seen rising in the Mediterranean and tropical seas, out of reach of
the fish-hunters of the water.

And now we must leave the open sea and steer for the
shore. It is true that many other fish are wandering in the
broad watery main, but many of them, such as the globe-fish,
feeding on the small crustaceans and the sea-horses,[36] whom
we shall meet nearer shore, are feeble forms carried hither
and thither by currents or on floating banks of seaweed, while
others have no special interest. The sharks, the mackerel,
and the flying-fish, are the most remarkable colonisers of the
ocean-surface, for even the enormous Sword-fish,[37] which
attacks the bonitos and whales with its long wedge-shaped

35   Dactylopterus.
36   Hippocampus.
37   Xiphias.

bony jaw, and is said to sail by raising his back-fin, is a distant off-shoot of the mackerel tribe.

So we cannot do better than follow our own common Mackerel, as they migrate in shoals out of the deep sea to feed on the fry of the herring or the pilchard in shallower water, or to leave their eggs floating not many miles from land, so that the tiny mackerel, when hatched, may live in the quiet bays till their strength comes.

But stop! Long before we have come so far as this, and while we are still a hundred miles or more from the shore, let us peep down into the sea-valleys, where forests of seaweed and marine plants are growing, and myriads of tiny sea-lice and crustaceans throng the water. What is that army of thin spindle-shaped forms rising and falling in such numbers? It is a shoal of herring, which have come there to feed upon the sea-animals, keeping out of sight of the sea-birds above, and the cod and sharks and ravenous fish which hunt them without mercy, so that they only venture to come to the surface on calm dark nights. It was in valleys such as these that the herrings were living when the older naturalists thought they were gone away to the Polar Seas, because they only saw them in spring and autumn, when they come into shallower water to drop their myriads of eggs,[38] which sink down, and stick to the seaweed and stones below.

But now they are revelling in the deep ocean, rising and falling with ease, for their air-bladder has two openings, one to the stomach and one to the outside of the body, so that the gas can adjust itself to their movements; and surely if the shark is the type of the old, lumbering, powerful, slow-breeding fish, the herring, with its narrow lissome body, light playful movements, and myriads of young, is the type of the

38   At least 10,000 for each mother.

new and active race. They are as truly social animals as any herds on land, for they travel in shoals of many hundreds of millions; and as they can squeak, and have a very good apparatus for hearing, it is more than likely that they call to each other. They make both the salt and fresh water their own; for when the eggs are hatched at the mouths of rivers the tiny fish take refuge there from the violent persecutions of the cod and mullet and haddock, flat-fish and whiting, and, together with the small fry of other fish, stroll up the rivers, where we call them "white-bait."

And now, as we come nearer to the shore, where countless numbers of small fry are filling the water, and all creatures are struggling together to accomplish three objects, namely, to get food, to avoid being turned into food, and to lay their eggs, we find many strange weapons and devices adopted by the different fish for protection and attack.

> ... "Each bay
> With fry innumerable swarms, and shoals
> Of fish, that with their fins and shining scales
> Glide under the green waves, in sculls that oft
> Bank the mid sea."

There are the Mullets,[39] with tender feelers under their chin, with which they brush the ground lightly as they swim, feeding on the tiny creatures. There are the walking fish, the Gurnards,[40] which have three of the spines of their arm-fins separate, and moved by strong muscles and nerves, so that they can walk on the sea-bottom, feeling their way, while the stiff, spiny rays of their back-fin stand up to wound any enemy attacking them from above. There are the tiny Blennies which walk

39   Mullus.
40   Trigla.

too, but by means of the few rays which alone remain
of their leg-fins growing close under the head. Then
there are the clinging-fish, the Gobies,[41] living on the
rocky shores, where the waves beat and roar, and they
have their leg-fins joined together, so as to form a kind
of funnel under their throat, with which they cling to
the rocks and then dart across the waves to feed, coming to anchor again out of the dash of the water; some
of these little fellows make nests and guard their eggs
after the mother has left them, till the young can shift
for themselves. More curious still, the Lumpsucker[42] has
its arm-fins and leg-fins all joined together into a round
disk under the throat, and so holds on bravely against the
dashing tide, defending the eggs which have been laid in
the seaweed near the shore, and even remaining to take
up the young ones when hatched, and carry them safely
back into deep water as they cling to his sides.

Meanwhile, close down upon the sand are the hiding-
fish, the Weevers, the Anglers, and the Flat-fish.

The weevers[43] are the most dangerous. Their shaded
yellow colour hides them from view, while the sharp spines
of their back-fins, which they keep raised, will inflict very
severe, if not poisonous, wounds on any creature striking against them. Nor is this all, for behind the cheeks,
fastened on to the horny gill cover, are daggers with
which they can strike, deliberately jerking them back so
as to give a sharp blow. These are fighting aggressive
fish, waging the war that goes on so sharply all round
our coasts.

41   Gobidæ.
42   Cyclopterus.
43   Trachinidæ.

Fig. 11.

The Fishing Frog.[44]

But there is one even more cunning than they, lying hidden in the seaweed or the sand—a large, flat, soft fish, about three feet in length, and quite half as broad as he is long, with a soft stumpy tail, stretching out behind, and a kind of wrist-joint to arm and leg fins, by which he can creep noiselessly along. His wide mouth is gaping open, so that a two-foot rule could be passed crossways into it, and his pointed teeth are bent back to allow his prey to enter. But how is this prey to be caught, for he is not going to move to fetch it? Notice all round his head and his body, the skin is fringed like blades of seaweed and plays about in the water; while above his head and back the spines of

44   Lophius piscatorius.

his fin stand up quite separate, and the front one is tapering and long like a fishing-rod, with a lappet at the end
like a bait. And now, as the shallow water ripples over his
head, the lappet plays to and fro, and the unwary fish come
up to nibble at it, lower and lower he waves it, and the
nibblers follow, till, opening his wide gape, he gulps them
down, even if they are as large as himself, and lies passive
with his swollen stomach till they are digested. This is
our own Fishing-frog,[45] of which one was once found with
seventy herrings in his stomach. He has relations all over
the world—in the open sea and down in its depths, and all
of them more or less follow his fishing habits. Yet there
is no creation of special parts for these strange weapons;
the altered back-fin and the jagged skin do all the work,
just as in some curious fish of the weever family in the
tropics, called the Stargazers,[46] the feelers on their lips,
longer than those of other fishes, and a lengthened thread
from below the tongue, play in the watery currents and
attract the small animals, while the fish with upturned
eyes watches them as they are lured to destruction.

Lastly, among all these curious forms upon our shores
there is an abundance of flat-fish—soles and turbot, brill
and plaice—flapping along at the bottom, covering themselves with sand, or rising up with that strange wavy
movement of the whole body in which they use what look
like long side-fins, but which are really the back-fin and
the belly-fin.

45   Lophius.
46   Uranoscopus.

Fig. 12.

The Common Sole.[47]
Above are two small soles as they swim when young. At that time they
are not larger than a grain of rice.—(*Adapted from Figuier and Malm.*)

If we wanted to pick out the strangest and strongest
proof of how the shape of fish is altered to suit their wants,
we need seek no further than the flat-fish.

When we were speaking of the shark order we saw that
the rays and skates are flattened forms suited to hide in the
sand, and these fish are truly spread out as if they had been
squeezed under a heavy weight, their broad arm-fins edging
the sides of their body. But the bony flat-fish, the Soles and
Turbot, have a far stranger history. The young sole, when
it comes out of the egg, is not flat like the young skate, but
a very thin spindle-shaped fish, something like a minnow.
He is then about the size of a grain of rice, very transparent,
and lives at the top of the sea. He has one eye on each side,
like other fish, only one eye is higher up than the other, and

47   Solea vulgaris.

the single fin on its back and the one under its body reach almost from head to tail. In this way he swims for about a week, but he is so thin and deep, and his fins are so small, that swimming edgeways is an effort, and soon he falls down on one side, generally the left, to the bottom of the sea. Many times he rises up again, especially at first, till he has got used to breathing at the muddy bottom, and meanwhile the eye that lies underneath is gradually working its way round to the upper side, his forehead wrinkles so as to draw the under eye up, while his whole head and mouth receive a twist which he never afterwards loses. His skeleton, it must be remembered, is still very soft, and the bones of his face are easily bent; and at last this eye is screwed round, and as he lies at the bottom he can look upwards with both eyes and save the under one from getting scratched by the sand, as it must have done if it had remained below.

Nor is this all, for while his under side, shaded from the sunlight, remains white and colourless, his upper side gradually becomes coloured like the sand in which he lies, and he is safely hidden from attack as he flaps along, feeding on worms and other animals. And now when he swims he no longer uses his arm and leg fins, which are quite small and insignificant, but bends his whole body, using the back and belly fins to help him. What we then call the top of the sole is really his side, where you may see the dark line of scales running along the middle, and one arm-fin lying close to his head. Yet he can swim strongly and to far distances, for in the winter the soles, too, migrate into the open sea, where they may be found in the deep water of the Silver Pit, between the Dogger Bank and the Well Bank. And now, before we leave the shore, we must glance at a curious weakly little fellow clinging by his curly tail to the seaweed, whom you will

Fig. 13.

Hippocampus, a fish commonly called the Sea-Horse.

certainly not take for a fish, even if you can find him out, so entangled is he generally in weeds of the same colour as himself. Yet the Sea-horse[48] is a true fish, covered not with scales but with plates, with which he makes a clicking noise by scraping them together. What look like large ears are really his arm-fins, while at the end of his long snout is a mouth shaped like an ordinary fish's mouth, but toothless, and he breathes with fish's gills arranged in round tufts instead of folds. What the use of his strange shape is to him we cannot tell, but at any rate his fleshless bony body must protect him from other fish, while his power of clinging causes him to be often carried by floating weed even into the open ocean, and make up for his feeble powers. In one thing he surpasses most other fishes, for he is a most careful father, carrying the mother's eggs in a little pouch under his body till the young ones escape. There is one form of these sea-horses in tropical seas which has long red fringes floating from its body, so that it cannot be distinguished from the seaweed in which it hides.

So we see that the deep sea, the open sea, and the shore,

48   Hippocampus.

are filled so full of different forms that there are enough not only to make use of every part, but also to provide food for each other, and we also see that by far the larger number even of widely-spread fish come near to the shore to leave their spawn, while the young ones often make their way into the brackish water at the mouths of rivers, and spend their youth in the shelter of the still fresh water.

Now it is very natural that many such fish should learn to remain in this quiet refuge, and in time to live there alto-gether. And because fish-life in the rivers is comparatively uneventful and little varied, we find much fewer peculiarities in river-fish. Many of them are very near relations of sea forms. There is the salmon, a true sea-fish, which wanders up the river to spawn in the pebbly shallows; and there are the trout, his near relations, which have learned to live entirely in the rivers. There are the sea-perches, large strong fish, and the smaller river perch, which have made their homes very successfully in the rivers, for their spines are so sharp that even the greedy pike hesitates to swallow them. There are the sea-sticklebacks, and the little river-stickleback.[49] This last is a very clever little fish, which hollows out the foundation of his nest very carefully in the bed of the river, and then builds it up for several inches with blades of grass and weeds (Fig. 14), gumming them together with the slime of his body. Then, when all is ready, he swims about to drive and coax the mother to the nest, sending her in to lay her eggs, and then driving her right through and out at the other side, so that a stream of water flows constantly over the eggs till they are hatched. Nay, his care does not end here, for when the young fish come out of the egg with a bag of yelk hanging under the body, as all young fish have

49   Gasterosteus.

Fig. 14.
STICKLEBACKS AND THEIR NEST. (Gasterosteus aculeatus.)

at first, and so cannot swim easily and escape their enemies, the courageous little father will defend them and fight fiercely with any fish which thinks to make a meal upon them, not leaving them till all the yelk is absorbed, and they are able to swim and feed themselves.

Besides these active river-fish there are the little stupid Miller's Thumbs,[50] hiding under the stones to feed on tiny animals; they are feeble relations of the gurnards which we saw walking on the bottom of the sea. Then there are the purely freshwater fish, the Pike and the large Carp family, with its many branches, the Roach, and Dace, and Gudgeon, and Minnow; and the enormous family of Cat-fish and Sheat-fish,[51] of which we have none in England, but plenty in America and other parts of the world, a family in which the fathers sometimes carry the eggs in their mouths till hatched. And last but not least among the freshwater forms is that irrepressible family of the Eels which we saw wandering in the deep sea, and which are also to be found near the shores all over the world. These fish will even travel through pipes and into cisterns; and will climb up trees so as to drop into neighbouring streams and continue their wanderings; they sleep in the mud in winter; and even after being frozen come to life again; and in the spring they go to the sea to spawn, giving rise to those shoals of young ones from three to five inches long which come in incredible numbers up the rivers in summer, making the eel-fairs,[52]—

"The silver eel, in shining volumes rolled,"

so much spoken of in old books, when the eels will often climb

---

50   Cottus.
51   Siluridæ.
52   More properly eel-fares (fare, Saxon, to travel; ex., way-faring man).

high banks, nay, even pass over miles of dry land, closing down their narrow gill-openings, and so shutting in water to serve them as they go.

All these, and many other freshwater families, show us how the fish have wandered into every possible nook of the waters, so that even in those inland salt lakes of North America and Asia into which no rivers flow fish-life is abundant; and we can only suppose that the eggs must have been carried by water-birds in their flight, or by gusts of wind, or have arrived there in ages long ago, before these lakes were cut off from the rest of the watery world.

Yet some few fish besides the eels have been known to travel over land to find watery "pastures new;" the Climbing Perch[53] of India and the Doras of Tropical America will both travel many miles when their own ponds are dried, the perch breathing by the help of a special apparatus, and the doras probably shutting water into its gills; for necessity, even in fishes, proves the "mother of invention," and in special works on fish you will find accounts of numberless strange devices and adaptations by which they manage to survive in the struggle for life.

And now, collecting together all we have learned, let us in conclusion form a rough picture of the history of the fish-world. All over our globe, from pole to pole, and from the Indian Ocean round to the east, back to the Indian Ocean again, is one vast world of waters, with inlets and land-locked seas bordering its margins, and rivers pouring into its depths. In the past ages of the world these rivers and coasts and inlets have varied innumerable times, but the great ocean-mother has always been there to bear the

53   Anabas.

increasingly-varied forms in her bosom, and to enable them to wander where best they could preserve life.

And so from their beginning, when they were probably as feeble as the lancelet, these earliest and simplest backboned animals with their two pair of limbs as yet very variable both in their position and shape, have been spreading far and wide over the watery three-quarters of the globe. We have seen how the enamel-scaled fish had their time of glory, but were not able to hold their ground, because they were not agile and fish-like enough to escape their foes; and how the sharks by their strength and boldness remain monarchs of the sea to the present day. Then we have seen that in old chalk seas the new and active race of bony fish appear in force; some like the herring and the carp, with air-bladders, which had openings like the enamel-scaled fish, and these can dart from heights to depths; while others had closed air-bladders, and these remain with most ease at one level, and can sometimes, if necessary, use the gas in their bladder for breathing, if they are oppressed with muddy water; and lastly, some, such as the dorado, have lost their air-bladder altogether, and gain in freedom of action what they lose in lightness and buoyancy. And during the ages that have passed since this bony race began, different branches each in their own way have thrown out curious weapons and developed strange organs to help them in the battle of life, so that now we have deep-sea fish carrying their own light; fish with distensible stomachs swallowing prey larger than themselves; fish with large air-bladders and long arm-fins springing out of their own element and floating in air; angling-fish, walking-fish, clinging-fish, and hiding-fish; and even those whose shape is distorted, like the sole, to enable them to hide and hunt in safety; while, when the sea is full, we find new varieties pressing their way into every river and tiny stream, and even overland into enclosed waters. Nay! when we descend into the recesses of the earth and visit the underground pools of the dark caverns of Kentucky, there we come upon fish which have found a refuge in eternal darkness, and have lost not only the power of sight but actually the eyes themselves.

And here we must leave them to go to higher vertebrate animals. Although but little is known of fish-life, a very small part even of that little has been given here, and yet we take leave of it with the feeling that its changes and chances are greater than we can ever thoroughly learn. How much pleasure these creatures have in their water-world it would be difficult for us to say; but since we find them playing together, hunting together, sporting in the warm sunshine, and diving and gambolling in the open sea, and sometimes even calling to one another, we cannot but think that life has great charms for them in spite of the many dangers surrounding them. And when, low though they are in the scale of life, we find them (though curiously enough always the fathers) carrying the eggs, building nests for them, and defending the young, we see that even here, in the very beginning of backboned life, we touch the root of true sympathy, the love of parent for child.

THE HOME OF THE EARLY AIR-BREATHERS

CHAPTER IV.

HOW THE BACKBONED ANIMALS PASS FROM
WATER-BREATHING TO AIR-BREATHING,
AND FIND THEIR WAY OUT UPON THE LAND.

S O the backboned animals, as fish, have peopled the seas
and rivers, and, as the ages have past on, have become

more and more fitted to their watery life, little dreaming of another and different life in the world of air above them. And yet in the same pond with the little stickleback, so busy building his nest, there is a creature which could tell him that it is possible to live in both worlds, if only you have the proper machinery to do it with.

It is clear that if the backboned animals were ever to live upon land, after they had begun their career in the water, there must have been some among them which learned gradually to give up water-breathing, and to make use of free air; and we shall not have far to seek for creatures which will help us to guess how they managed it.

From almost every country pond, or ditch, or swamp, a chorus of voices rises up in the springtime of the year, calling to us to come and learn how Life has taught her children to pass from the water to the air; for it is then that the frogs lay their eggs, and every tadpole which grows up into a frog carries us through the wonderful history of an animal beginning life as a fish with water-breathing gills, and ending it as a four-legged animal with air-breathing lungs.

Come with me, then, to some stagnant pool in a country lane, towards the end of March, and there we shall no doubt find a whole company of frogs, croaking to their hearts' content after their long winter sleep in the mud at the bottom of the pond. They are wide awake now, and are actively employed laying their eggs. Look carefully around the edges of the pond, especially in that part where the wind has driven the scum to the side, and you will doubtless find in some still corner a gluey mass (*e*, Fig. 15), which looks like a lump of jelly with dark specks in it. Take this up carefully, for it is frog spawn; carry it home together with some weeds from the pond; put it in a glass bowl with water; and then from day to day you may study the history of a frog's life.

That jelly-like mass is a collection of frog's eggs. When

Fig. 15.

Metamorphosis of the Frog.

e. Eggs. 1. Tadpoles just out of the egg. 2. With outside gills. 3. With gills hidden, and beak-like mouth. 4. Hind legs appearing. 5. All legs grown, but fish-tail remaining. 6. Putting on Frog appearance; tail being absorbed. 7. Young perfect Frog.

they were laid, each egg was a small round dark body in a gluey covering, and they all fell to the bottom of the pond, where, by degrees, the water oozing through the envelope swelled each egg, till they clung altogether in a mass, and, rising, floated at the top. Then very soon each round dot lengthened out into a long streak, and in a few days an eyeless head appeared at one end with a soft closed mouth under it, and at the other a tail, with a soft fin round it like the tail of the lancelet; so that by the time you find the spawn, you may, most likely, be able to see the tiny creature wriggling every now and then in its watery bed. This will go on for some time, and a week or two may pass before the moving tadpole breaks through its egg skin, and coming out into the world, fastens on to a piece of weed (1, Fig. 15) by two little suckers behind its mouth. And now that it is out of the egg the interest begins. Look carefully day after day and you will see some branching tufts (2, Fig. 15) growing larger and larger on each side of its head. What are these? We have not seen them in any fish. No! but if you take a young hound-shark out of his leathery egg before his time, you will find that he has outside gills much like these, only he loses them before he comes out into the world, whereas the tadpole keeps them to breathe with a little longer. If you put the tadpole, at this stage, under the microscope, you can see the red blood flowing through these gills to take up air out of the water.

Meanwhile the tadpole's lips are gradually forming into a round mouth, much like the lamprey's, and by-and-by the inner part of this mouth is covered with two little horny jaws, forming a sharp beak (3, Fig. 15) with which he will nip off pieces of weed for food. Meanwhile, as he grows larger and larger, and eyes, nostrils, and flat ears form in the head, a covering begins to grow back over the sides of the neck, and little by little the branching tufts disappear (3, Fig. 15). How, then, can he breathe now? Watch carefully and you will see that he gulps every moment as we saw the minnow doing (p. 26). The fact is that the outside tufts have faded away, and under the cover the tadpole has six slits in his throat, like the slits of the lamprey, which are covered in somewhat similar fashion to those of the amphioxus (see p. 14), and he breathes through them.

Here is our tadpole, then, to all intents and purposes a fish. He swims with a fish's tail; he gulps in water at his mouth, passing it out at the slits in his throat after it has poured over his fish's gills. Moreover, he has a fish's heart, of two chambers only, like the minnow's (p. 26), which pumps the blood into these gills to be freshened, while, like the lamprey, he has a gristly cord, enlarged at the end to form a gristly skull, a round sucking mouth, and no limbs. All this time, however, though he has a fish's fin round his tail, he has no arm or leg fins. Wait a while and you will see that under his tender skin far more useful limbs are being prepared. As he grows bigger and more active week by week, wriggling among the weeds and feeding greedily, two little bumps appear one on each side of his now bulky body, just where it joins the tail. These bumps grow larger every day, until, lo! some morning they have pierced through the skin, and two tiny hind legs (4, Fig. 15) are working between the body and the tail. The two front legs are longer in coming, for they are hidden under the cover which grew over the gills, but in about another week they too appear, and we have a small four-legged animal with a lamprey's tail (5, Fig. 15). These legs are something far in advance of fish fins, for they have shoulders and thighs, arm and leg bones, wrist and ankle bones, hand and foot bones; and instead of the large number of rays in a fish's fin they have four fingers on their short front legs, and five toes at the end of long hind ones; the toes being joined together by a web, which helps him wonderfully in striking the water as he swims.

The tadpole has now become fitted to jump and leap on the land or swim by his legs in the water; and, moreover, while these legs have been growing, another change has been taking place. You will notice by careful watching that at first he still gulps in water as he used to do, but he comes more often to the top, and, poising himself so that his mouth is out of the water, gives out a bubble of bad air, draws in some fresh, and goes down again. Why does he do this? Have you any recollection of another fish-like animal which comes up to take in air? Look back at our friends the mud-fishes (p. 36), and read how the Ceratodus fills his air-bladder when he is short of good air in the water. When you have re-read this, you will suspect that the tadpole, too, has something like an air-bladder, which he fills from time to time. And so he has. While his legs are growing a bag has been forming inside at the back of his throat, which afterwards divides into two, and he fills these by shutting his mouth, drawing air in at his nostrils, putting up the back of his tongue to shut it in, and then swallowing it down into the lungs; so that he is now a truly double-breathing animal, using his gills when below water and his lungs when above. Moreover, if you could watch inside his body, you would now see that little by little the blood-vessels going to the gills grow smaller and smaller, and those going to the lungs grow larger and larger; while the fish's two-chambered heart divides into three chambers, one to receive the blood from the body, another to receive it from the lungs, and one to drive this blood back again through the whole animal. And when at last this change is so complete that all the blood goes to the lungs to be freshened, the gills shrivel up and disappear, and our tadpole is a true air-breathing animal.

Notice, though, that he is still cold and clammy, not warm like a mouse or a bird. For his blood still moves slowly, and as he has only three chambers to his heart instead of four, as warm-blooded animals have, the good blood from the lungs and the worn-out blood from his body become mixed each time they come round, so that his breathing work is still of a low kind all his life. And now that he can leap and swim with his legs, his tail is no longer of use to him, and it is gradually sucked in, growing shorter and shorter till it disappears, and the young frog is complete.

Thus our backboned animal has succeeded in getting out of the water on to the land, and in doing so he has quite changed his habits. A peaceful vegetarian before, he is now a greedy eater of insects, slugs, and other animals. His horny beak has been pushed off; his lips have stretched back farther and farther, till they now open right back as far as his flat little ear; and he is a gaping, wide-mouthed, leaping frog[54]—

> ... "Hoarse minstrel of a strain
> Aquatic, leaping lover of the rain;"

(7, Fig. 15), with teeth in the roof of his mouth. But perhaps his tongue is the most curious of all, for instead of being fixed at the back, and free in the front, as in most other animals, the root of it is fastened to the front of his lower jaw, and the tip lies back in his mouth, so that when he wishes to catch an insect he throws his tongue quickly

---

54   Not "waddling;" it is the toad, not the frog, that waddles.

forward, captures his prey on the sticky point, and flings it back down his throat.

So he hops about the summer long, if he can only escape from ducks and rats and other frog-eating animals. He often takes to the water, for he can fill his lungs with air and use it very slowly, and, moreover, his soft skin is of great use to him in still breathing in the water or in the moist air; and when winter comes he takes refuge with many others at the bottom of the pond, and sinks into a state of torpor, till the spring brings croaking and egg-laying time round again.

Fig. 16.

The Common Smooth Newt[55]—male and young
in the water; female on the bank.

Our little frog, then, is truly an animal with a double

55　　　Lissotriton punctatus.

life, a genuine amphibian,[56] meaning by this, not merely
an animal that can swim in the water and move on land,
for seals and water-rats, white bears and hippopotamuses,
can do this, but one that in the early part of its life would
die if taken out of the water, while afterwards it lives and
breathes in the air.

Fig. 17.

Proteus of the Carniola caverns,[57] with its external
breathing gills.—(*ADAPTED FROM BREHM.*)

Have these double-lived creatures, then, such a great
advantage over real water animals, or how can we account
for their having adopted this strange life? If we only look
upon them as they are now, we can scarcely call them par-
ticularly successful, compared to other animals. For though
there are plenty of them, yet they are comparatively small and
insignificant; and when we find large ones like the gigantic

56   *Amphi*, all around; *bios*, life.
57   Proteus anguineus.

salamander of Japan, they are sluggish and feeble. Look at the common newts, or water-salamanders of our ponds, with their weak crawling limbs, as they wander round the edges of a pond, feeding on water-insects and tadpoles, the male with his crested back, the smooth mother, and the young eft-tadpole with its branching tufted gills (Fig. 16). They are much less active than the frog, for they never lose their tails, and they come less often out of the water, although they are true air-breathing animals. Then, when we go to other countries, there is the Proteus (Fig. 17), that curious half-transparent newt, with a round body and tiny helpless legs, which lives in eternal darkness in the still underground pools of the Carniola caverns near Adelsberg. He has become well fitted for his dismal life, for his tiny eyes are grown over with skin, and he never loses the feathery gills on each side of his neck, but lives like a tadpole all his life, although he has true lungs. Again, in America we have the Siren, with its long snake-like body, and only front legs, with which it cannot walk. It, too, keeps its gills as it wanders about the stagnant waters of South Carolina, feeding on worms and insects. Then in the Mexican lakes there are the curious Axolotls, which also wear outside gills, as a rule, all their lives, and fathers, mothers, and children remain breathing in the water together, although they have real lungs. But about twenty years ago, some of those axolotls, which were kept in the Jardin des Plantes in Paris lost their gills, came out upon the land, and astonished people by becoming true land salamanders, like some already well known and called Amblystomes, breathing only with their lungs. It was difficult for some time to make the world believe that grown-up water-breathing creatures which could lay eggs were able to turn into other creatures without gills. But at last a lady, Fraulein Marie von Chauvin, took some axolotls when they were full-grown, and kept them on land in wet moss, washing and feeding them every day, and thus succeeded in teaching them to breathe air, so that their gills shrivelled up and disappeared. Then there could no longer be any doubt that the axolotl is only the lower water-form of the amblystoma, which in the Mexican lakes, owing to the increased dryness of the surrounding country, has lost the habit of coming out on to the land, and remains in the water

Fig. 18.

Axolotl, a creature living and breeding for generations in the water. Amblystoma coming out of the water,—an axolotl which has lost the gills and acquired lungs.

with its little ones all its life; but which, when brought to a moist climate where it can breathe comfortably on land, sometimes returns to its old double life.

We have, in fact, in Europe real land salamanders, which live in cool damp places, looking like lumpy soft-skinned lizards, but going down to the water to lay their eggs, that their little ones may go through their tadpole life—and one of these, the black salamander,[58] which lives high up in the mountains of Germany, France, and Switzerland, does not even go to the water, but carries the young tadpoles in her body till they can breathe air and run alone; and yet they are still true *amphibia*, for if they are taken out of their mother and put in water, they go through all their changes like common efts and newts.

Lastly, there is a strange group of legless creatures called Cæcilians, which have taken refuge underground, burrowing like worms, though they are true amphibians and their young have gills in their babyhood hidden under a slit in the neck. These cæcilians are the only amphibians which have scales something like fishes, yet they never live in the water, but in the marshy ground of tropical countries, feeding on worms and insects.

* * * * *

Now when we think that these sluggish newts, and salamanders, and cæcilians, with their more nimble but comparatively unprotected relations, the frogs, are all the amphibians now living, we cannot but wonder how Life came to produce such a feeble set of creatures to fight the battle of existence.

But if we glance back to that far-off time when the ancient fishes were wandering round the shores and in the streams of the coal-forests, we shall be better able to read the riddle.

58   Salamandra atra.

For in those days it was a great step for an animal to get out of the water at all, and those that did so had a much better time of it than our frogs and newts have now, when the country is full of land enemies.

And so we find that the *amphibia* were not then the small scattered groups they are now, but strong lusty animals, with formidable weapons. In the hardened mud, which in those days formed the soft swampy ground of the coal-forests, but is now stiffened into the roofs and floors of our coal-mines, footprints have been left which tell us of large and formidable creeping animals, with toed feet and long flat tails, dragging themselves over the marshes of the coal-forests, and finding their way to many places which even the mud-fish with their paddles could not reach; and from time to time, in these same roofs and floors of our mines, both here and in America, we find the bones and coverings of these *amphibia*, buried in Nature's catacombs for ages, and only brought to light by the rude hand of man.

These remains remind us that

> "A monstrous eft was of old the lord and master of earth,
> For him did the high sun flame, and his river billowing
> ran.
> And he felt himself in his force to be Nature's crowning
> race;"

for they show us huge and powerful creatures[59] which sported in the water or wandered over the land with sprawling limbs, long tails, and bones on which gills grew, while their heads were covered with hard bony plates, and their teeth were large, with folds of hard enamel on the surface. Some of these were fish-like, with short necks and broad flat tails, but they had true legs and toes; others, more like crocodiles, and

---

59   See Picture-heading, p. 71.

sometimes ten feet long, were able to walk firmly, but still dragging their bodies and long tails over the swampy ground on which their footprints are still found; some were small and more like lizards, with simple teeth, scaly armour, and light nimble bodies; and these, probably, ran about quickly on the land, and have sometimes left their skeletons in the hollow trunks of the old coal-forest trees.

All these plated and formidable creatures were *amphibia* or double-lived animals, and this was *their* Golden Age, as they preyed upon the fishes in the swamps and ponds, probably not sparing even their nearest connections, the mud-fishes, who, less fortunate than themselves, had followed the road of fish-life instead of coming out upon the land. They lived so long ago that we can tell but little of their daily lives, but it is clear that they played a very different part from our small frogs and newts of to-day, and in their well-formed limbs were worthy forerunners of land and air-breathing animals.

But like the old race of fishes these large amphibians were only to have their day, for as other branches of the family tree grew up, and reptiles grew strong and mighty, and other true land animals began to flourish, these huge plated forms dwindled away, and we lose sight of them; and when we find any of their relations again it is only as our present frogs and newts, salamanders and cæcilians, which have taken up their refuge in lakes, ponds, ditches, underground waters, or damp mud. And, curiously enough, those forms of to-day which are most like the huge *Labyrinthodonts*,[60] as they are called, of the old coal-forests, are the feeble cæcilians, with their horny scales and their numerous ribs, although they have now fallen the lowest of all amphibians, and, with their

60   Labyrinthodonts (*Laburinthos*, spiral; *odontas*, teeth).

sightless eyes and ringed and legless bodies, have taken to burrowing in the ground like worms.

Not so the frogs, which, like the bony fishes, began their career in later times, and have known how to fit themselves into many nooks and corners in life. In almost all countries of the globe they hop merrily about the ponds and ditches, never wandering far from the water, into which they jump and dive whenever danger threatens. It is true they are eaten by thousands, both as tadpoles and frogs, by birds, snakes, water-rats, and fish, and even by each other, but they multiply fast enough to keep up the supply, and find plenty of insects both in and out of the ponds. Nor have they kept entirely to a watery life, for their near relations, the toads, which have toothless mouths and toes less webbed, have ventured much farther on to the land, protected partly, no doubt, by the disagreeable acrid juice which they can throw out from a gland behind the eye whenever they are attacked.

It is curious to notice the quiet leisurely waddle of the sluggish toad, as he spreads out his short fat legs and puffs out his warty skin, and to compare him with the nervous, anxious, little frog, starting at every danger. And still more curious is it to see him getting out of his skin, as he does several times a year. For his skin does not peel off in pieces as it does in the watery frogs, but splits along his back; then he wriggles about till it lies in folds on his sides and hips, and, putting one of his hind feet between the front ones, draws the skin off the leg like a stocking off a foot. With the other leg he does the same, and then, drawing out his front legs, pulls the whole skin forward, and stripping it over his head, swallows it; thus deliberately putting his old coat inside him, and appearing in one that is glossy, fresh, and new. The toad has many enemies in spite of his acrid taste, and he shows his wisdom by hiding in walls and under stones in the daytime, and coming out in the dusk of evening to hunt the beetles and grubs so often out of reach of the water-loving frog.

Fig. 19.

The Flying Tree-Frog of New Guinea[61] (*WALLACE*).

But the toad is not the only land relation of the frog; there are others of the group that venture even farther

61   Rhacophorus Rheinhardii.

from water; for in most parts of the world (though not in England), tree-frogs, with sucking disks at the ends of their toes and fingers, climb the trees and hunt for insects among the leaves and branches; while in Borneo Mr. Wallace found one (Fig. 19) with webbed feet, which it spread out, and so flew down from the trees. There are plenty of the ordinary tree-climbing frogs to be seen in the south of France, their small green bodies peeping out from under the dull gray olive-leaves; and to be heard, too, in an endless chorus all night long when the spring arrives.

But how can these tree-dwellers bring up their little ones in water? Some of them come down and lay their eggs in the ponds, and even sleep down in the mud in winter. Others lay their eggs in little puddles of water in the hollows of the trees, and there the young ones live their tadpole life; while in one curious tree-frog of Mexico, called the *Nototrema*, the mother has a pouch in her back, and the father places the eggs in it for the little tadpoles to live in a moist home till they leap out as perfect frogs.

Nor is this the only case in which fathers and mothers take care of their young. In one species of frogs living near Paris, the father[62] winds the long string of gluey eggs round his thighs, and buries himself in the ground till the young tadpoles are ready to come out, and then he leaps into the water. And in one of the tongueless toads, the Surinam toad,[63] the mother's soft skin swells up, forming ridges and hollows, and when her eggs are laid the father clasps them in his feet, and, leaping on her back, puts an egg into each hollow. Then the mother goes into the water, and remains

62   Alytes obstetricus.
63   Pipa Americana.

there while each tadpole completes its changes in its own hole, jumping out at last a finished toad.

Yet, in spite of curious habits such as these, the frogs and their companions on the whole lead a very monotonous life. They are, it is true, more intelligent than fish, and have learned to know more of the world, but in the long ages that have passed since their ancestors roamed in the coal-forest marshes, other and higher animals have taken possession of the land, and left room only for a few scattered groups of *amphibia*. Still, however, they remain hovering between two lives, and filling such spots as neither the fishes nor the land animals can occupy; and when we hear them croaking in the quiet night, or see them leaping on the marshy ground, they remind us that we have still living in our day, a link between the fish whose world is a world of waters, and the air-breathing animals which have become masters of the land.

THE REPTILES IN THEIR PALMY DAYS

CHAPTER V.

THE COLD-BLOODED AIR-BREATHERS OF THE GLOBE IN TIMES BOTH PAST AND PRESENT.

AND now the transformation is complete, for when we pass on to the next division of backboned animals, the "Reptiles," we hear nothing more of gills, nor air taken from the water, nor fins, nor fishes' tails. From this time onward all the animals we shall study live with their heads in the air, even if their bodies may be in the water; they swim with their legs or, as in the case of the snakes, with their wriggling bodies, and they lay their eggs on the land where their young begin life at once as air-breathers.

Yet they can often remain for a long time both under water and under ground, for they are still cold-blooded animals, breathing very slowly, and easily falling into a state of torpor when the air around them is cold and chill. They are but the first step, as it were, to active land-animals; yet they have played a great part in the world, and when we know their history we shall be surprised to find how much Life has been able to make of her cold-blooded children.

To learn how this has been, however, we must travel away from home and our own surroundings. The tiny brown lizard which runs over our heaths, while its legless relation, the slowworm, burrows in the ground,—the few snakes which glide through the grass of our meadows, and the stray turtles thrown at rare intervals on our shores,— tell us very little about true reptile life. It is to Africa, India, South America, and other warm countries, that we must go to find the formidable crocodiles, huge tortoises, large monitor-lizards, and dangerous boa-constrictors, cobras, and rattle-snakes. And even then, strong and powerful as some of these creatures are, they do not tell us half the history of the cold-blooded air-breathers. For the day of reptile greatness, like that of the sharks and enamel-scaled fish, was long long ago.

Now that we know how frogs pass from water-breathing to air-breathing, and how axolotls, accustomed to live all their life in the water, can lose their gills and become land-animals, we can form an idea how in those ancient days, while still the huge-plated newts were wandering in the marshes, some creatures which had lost their gills would take to the land, and their young ones starting at once as air-breathers, as the black salamanders do now (see p. 80), would in time lose all traces of the double or amphibian life, and become true air-breathing reptiles.

At any rate, there we find them appearing soon after the coal-forest period passed away, at first few and far between, in company with the large amphibians, but spreading more and more as the ages passed on, till they in their turn became monarchs of the globe. Already, when the coal-forests had but just passed away, a lizard,[64] in some points like the monitors that now wander on the banks of the Nile, was living among his humbler neighbours; and from that time onwards we find more and more reptiles, till just before the time when our white chalk was being formed by the tiny slime-animals at the bottom of the sea, we should have seen strange sights if we could have been upon the globe. For the great eft was no longer

"... lord and master of earth."

All over the world, and even in our own little England, which was then part of a great continent, cold-blooded reptiles of all sizes, from lizards a few inches long to monsters measuring fifty or sixty feet from head to tail, swarmed upon the land, in the water, and in the air. There were among them a few kinds something like our tortoises, lizards, and crocodiles; but the greater number were forms which have

64   Protorosaurus or Thuringian lizard.

quite died out since birds and beasts have spread over the
earth, and a wonderful and powerful set they were.
Some were vegetable-feeders, which browsed upon
the trees or fed upon the water-weeds, as our elephants
and giraffes, our hippopotamuses and sea-cows do now.
Others were ferocious animal-eaters, and their large
pointed teeth made havoc among their reptile companions,
as lions and tigers do among beasts. Some swam in the
water devouring the fish, while others, like birds or bats,
soared in the air.

In the open ocean were the sea-lizards, some called
Fish-Lizards,[65] like huge porpoises thirty feet long, but
really cold-blooded reptiles, with paddles for legs, and
long flattened tails for swimming. Woe to the heavily-
enamel-scaled fish when these monsters came along, their
pointed teeth hanging in their widely-gaping mouths as
they raised their huge heads, with large open eyes, out
of the water! Then among these were others with long
swan-like necks and small heads,[66] which would strike
at the fish below them in the water, while other slender,
long-bodied monsters,[67] measuring more than seventy feet
from tip to tail, flapped along the sea-shore with their
four large paddles, or swam out to sea like veritable sea-
serpents, devouring all that came in their way. These were
all water-reptiles, while there were also many smaller
land-lizards playing about upon the shore, and among the
trees and bushes. But the strangest of all were perhaps
the "Flying reptiles"[68] of all sizes, from one as small as a
sparrow to one which measured twenty-five feet from tip

65   Ichthyosaurus.
66   Plesiosaurus.
67   Mosasaurus and Clidastes.
68   Pterodactyls.

to tip of its wings. These reptiles did not fly like birds, for they had no feathers, but only a broad membrane, stretching from the fifth finger of their front claw to their body, and with this they must have flown much as bats do now, while some of them were armed not only with claws, but also with hooked beaks and sharp teeth, with which they could tear their prey.

And meanwhile upon the land were wandering huge creatures, larger than any animal now living, which were true reptiles with teeth in their mouths, yet they walked on their hind legs like birds, probably only touching the ground with their short front feet from time to time, as kangaroos do. They had strong feet with claws, the marks of which they have left in the ground over which they wandered, supporting themselves by their powerful tails as they went.

Some of them were peaceful vegetarians,[69] browsing on the tree-ferns and palms, and rearing their huge bodies to tear the leaves from the tall pine-trees. But others were fierce animal-feeders. Fancy a monster thirty feet high,[70] with a head four or five feet long, and a mouth armed with sabre-like teeth, standing upon its hind legs and attacking other creatures smaller than itself, or preying upon those other huge reptiles which were feeding peacefully among the trees. Surely a battle between a lion and an elephant now would count as nothing compared to the reptile-fights which must have taken place on those vast American lands of the west, or on the European pasture-grounds, where now the remains of these monsters are found.

69   Iguanodon in Europe, Hadrosaurus in America.
70   Megalosaurus in Europe, Dryptosaurus in America.

But where are they all gone? We know that they have lived, for we can put together the huge joints of their backbones, restore their gigantic limbs, and measure their formidable teeth, but they themselves have vanished like a dream. As time went on, other and more modern forms, the ancestors of our tortoises, lizards, crocodiles, and afterwards snakes, began to take the place of these gigantic types; while warm-blooded animals, birds and beasts, began to increase upon the earth. Whether it was that food became scarce for these enormous reptiles, or whether the birds and beasts drove them from their haunts, we are not yet able to find out. At any rate they disappeared, as the ancient enamelled fishes and large newts had disappeared before them, and soon after the beds of white chalk were formed, which now border the south of England and north of France, only the four divisions of tortoises, lizards, crocodiles, and snakes, survived as remnants of the great army of reptiles which once covered the earth.

* * * * *

Ah! if we could only have a whole book upon reptiles to show how strangely different these four remaining groups have become during the long ages that they have been using different means of defence; and how, even in a single group, they employ so many varied stratagems to survive in the battle of life! Look at the tortoises with their hard impregnable shells, the crocodiles with their sharp-pointed teeth and tough armour-plated skins, and the silently-gliding snakes with their poisonous fangs or powerful crushing coils. See how the tiny-scaled lizard darts out upon an insect and is gone in the twinkling of an eye, and then watch the solemn chamæleon trusting to his dusky colour for protection, and

scarcely putting one foot before another in the space of a
minute.

Each of these has his own special device for escaping
the dangers of life and attacking other animals, and yet we
shall find, before we finish this chapter, that they are all
formed on one plan, and that it is in adapting themselves
to their different positions in life that they have become so
unlike each other.

We shall all allow that the Tortoises are the most singular
of any, and it is curious that they are also in many ways
the nearest to the frogs and newts, although they are true
reptiles. Slow ponderous creatures, with hard bony heads
(Fig. 20), wide-open expressionless eyes, horny beaks, and
thick clumsy legs, the tortoises seem at first sight to be only
half alive, as they lumber along,

> "Moving their feet in a deliberate measure
> Over the turf,"

carrying their heavy shell, and eating, when they do eat, in
a dull listless kind of way. They do, in truth, live very feebly,
for they can only fill their lungs with air by taking it in at
the nostrils and swallowing it as frogs do, and then letting
it drift out again as the lungs collapse, for their hard shell
prevents them from pumping it in and out by the movement
of their ribs like other reptiles. This slowness of breathing
and the fact that they have only three-chambered hearts
like frogs (see p. 77), so that the good and bad blood mix
at every round, causes them to be very inactive, and they
digest their food very slowly, and have been known to live
months and even years without eating.

This sluggishness would, indeed, certainly be their ruin
in a bustling greedy world, if it were not for the strong

Fig. 20.

The Greek Tortoise.

box in which they live. Take in your hand one of the small Greek[71] or American[72] tortoises, so often sold as pets, and you will see how well he can draw back out of harm's way, while at the same time you will, I think, be sorely puzzled to understand how he is made. His head, his four legs, and his tail, with their thick scaly skin, are intelligible enough. But why do all these grow on to the inside of his shell, so that when you trace them up you cannot find the rest of his soft body? You would hardly guess that his shell *is* the rest of his body, or at least of his skeleton. But it is so. The arched dome which covers his back is made of his backbone and ribs,

71   Testudo Græca.
72   Testudo talenlata.

and the shelly plates arranged over it are his skin hardened into horny shields, which, in the Hawksbill turtle, form the tortoise-shell which is peeled off for our use; while the flat shell under his body is the hardened skin of his belly, and the bones which belong to it.

Let us make this clear, for it is a strange history. If you look at the skeleton of a lizard (Fig. 23, p. 103), it is all straight-forward enough. His head fits on to his long-jointed backbone, which is able to bend in all parts freely, down to the very tip of his tail. His front legs with their shoulder

Fig. 21.

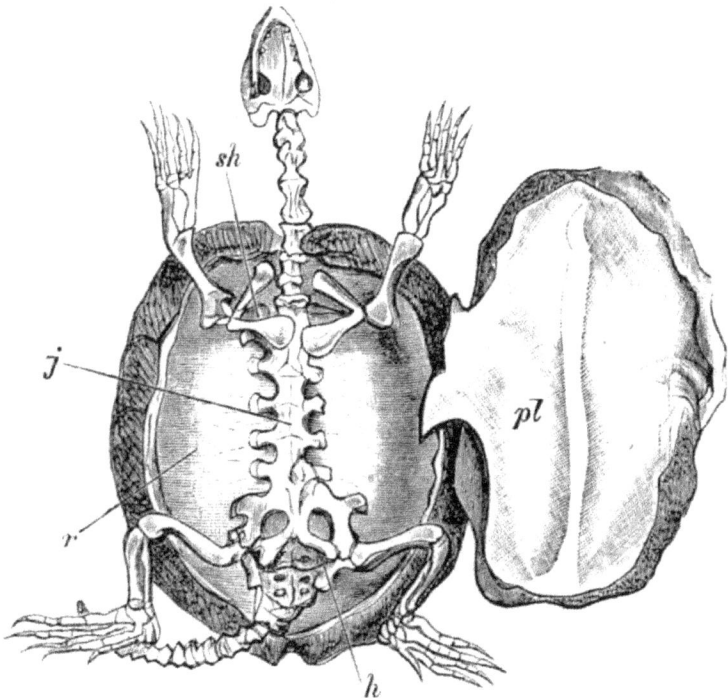

Carapace of the Tortoise.
j, Joints of the backbone grown together; *r*, ribs formed into a solid cover; *sh*, shoulder bones; *h*, hip bones covered by carapace, which has grown over them.

bones (s), and his hind legs with their hip bones (h), are
attached in their proper places to his backbone, and lastly,
his ribs (r) protect the inside of his body, and by expand-
ing and contracting pump the air in and out of his lungs,
the front ribs being joined underneath in a breastbone. It
is easy to see, therefore, that the lizard may be active and
nimble, twisting his body hither and thither, and escaping
his enemies by his quickness. But the tortoise is slow and
sluggish, and has only managed to baffle the numberless
animals which are looking out for a meal by fabricating a
strong box to live in. But he had to make this out of the same
kind of skeleton as the lizard, with the one difference that
he has no breastbone. Let us see how it has been brought
about. The bones of his neck are jointed and free enough
as you can see (Fig. 21), and so are the joints of his tail,
beginning from behind his hip bones (h). But with his back
it is different. The backbone can be clearly seen inside the
empty shell, running from head to tail so as to cover the
nerve-telegraph, but the joints (j) have all grown together,
and on the top they have become flattened into hard plates,[73]
while the ribs (r) which are joined to them have also been
flattened out and have grown firmly together so as to make
an arched cover or *carapace*. If now you look at the back of
the young tortoise (Fig. 22), which has been taken out of
the egg before it was full-grown, you will see these plates
(p) on the side where the tortoise-shell (ts) has been peeled
off. They have not yet widened out enough to be joined
together, and the ribs (r) are as yet only united by strong
gristle. But what is that row of oblong plates (mp) round
the edge? Those are the marginal plates, and they are mere

---

[73]   The parts of the joints which flatten out in the tortoise are seen at sp in the
lizard and snake, pp. 103, 110.

Fig. 22.

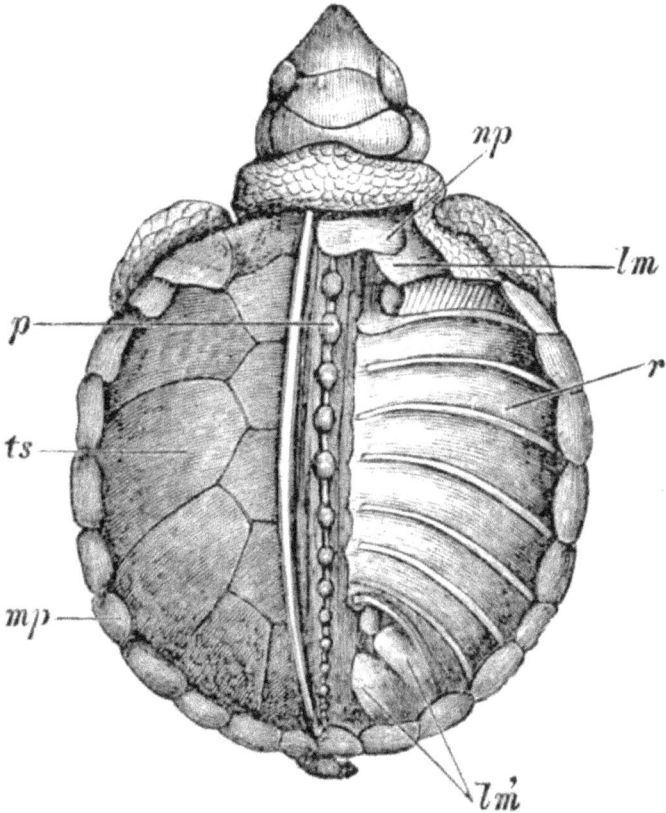

Back of a Young Tortoise.—(*From Rathke.*)
ts, Tortoise-shell covering the whole carapace; this has been removed on
the right side; *mp*, marginal plates binding the edges of the ribs; *np*, neck-
plate; *p*, plates formed of the top of the backbone joints which have grown
together; *r*, ribs which have not yet spread out so as to form a continuous
shell; *lm*, *lm'*, front and hind leg muscles not yet covered by the carapace.

skin bones, like the bony plates of the crocodile, but they
are all firmly fixed together so as to bind the edges of the
ribs, while plates of the same kind form the shell under the
body, and the whole is covered by the horny skin.

But there still remains another great puzzle. How come

the shoulder bones and hip bones of the tortoise to be inside his ribs instead of being outside them, as in other animals? But look again at our baby tortoise, and you will see that the muscles of his front legs (*lm*, Fig. 22) are not covered by ribs, neither are those of his hind legs (*lm'*). They stand just like those of other animals, in front between the ribs and the neck, and behind between the ribs and the tail. But as the tortoise grows up, the bony plates press forwards and backwards, and cover up the shoulders and hips, protecting the soft legs and neck, and giving him the curious appearance of living inside his own backbone and ribs.

In this way, then, the tortoises have managed to hold their own in the world. Living slowly, so that they sometimes go on growing up to eighty years old, wanting but little food, and escaping the cold by sleeping the winter months away in some sheltered nook, they ask but little from Life, while they escape the dangers of sluggishness by growing their skeletons so as to form a citadel which even birds and beasts of prey can rarely break through. They are, it is true, often eaten when young, and the jaguar of Brazil knows how to dig the poor American tortoise out of his shell and eat him; while large birds are formidable enemies to our Greek tortoise, and are said to drop it down on the rocks, and break it to pieces. But, on the whole, they escape most of these dangers, and wander in the woods and dry sandy places of sunny Greece and Palestine, laying their bullet-shaped eggs in warm spots to hatch, seldom wandering far from home, and lying down for their winter's sleep under heaps of drifted leaves or in holes of the ground.

These are true Land-tortoises,[74] and so are the gigantic tortoises which used to live in the island of Aldabra, and

74   Testudinea.

others still surviving in the Galapagos and other islands near Madagascar, which weigh at least 200 pounds, and on whose backs Mr. Darwin rode when he found them travelling up the island to get water to drink, feeding on the juicy cactus as they went. Some carapaces in our museums belonging to these tortoises measure four feet long and three broad; yet they were timid fellows when alive, drawing back completely within their shells when danger was near. We even find some smaller land-tortoises[75] in America, called the Box-tortoises, which have soft joints in their under shell, so that they can draw it up both in front and behind, shutting themselves completely in.

Not so the River-tortoises,[76] which are greedy animal-feeders, and as they live in the water do not need the same protection. Their box is much flatter and more open at the ends, so as to allow them to swim freely with their webbed feet; and they are fierce and bold, the Snapping Turtle[77] of the lakes and rivers of America being a terrible fellow, tearing the frogs and fishes in the water with his sharp claws, and even snapping strong sticks in half with his powerful beak. The Mud-tortoises, too, which swim swiftly with their strong legs and long neck outstretched, do not need a hard shell, and they have scarcely any plate below, and only a gristly leathery covering above, which looks very like the mud in which they hide.

Lastly the Sea-tortoises or Turtles, which swim in the warm parts of the Atlantic and Pacific Oceans, have only an open flat shell under which they cannot draw their head and feet, for they strike out boldly into the open ocean, feed-

75   Terrapins.
76   Emyx and Trionys.
77   Chelydra serpentina.

ing on seaweed, jelly-fish, and cuttle-fish, rowing grandly
along with their broad paddles which they feather like oars
as they go. They have only one time of weakness—when
they come on islands, such as Ascension and the Bahama
Islands, which they choose probably because they find fewer
large animals there. There the mother turtle arrives at night,
looking fearfully around, and if all is still comes flapping in
over the sand, and, clearing a hole with her flippers, lays
about 200 soft round eggs and covers them up and leaves
them. Then in about a month the young turtles come out
and make at once for sea, though many of them fall victims
to large birds of prey on their way. Woe, too, to the mother
when she is laying her eggs, if these large birds are near,
for she cannot defend her soft body; or, worse still, if the
natives are on the look-out; for then the Green Turtle,[78]
coming ashore from the Atlantic, is tilted over on her back
and killed for food; and the Hawk's-bill Turtle[79] from the
Indian or Pacific Oceans is cruelly stripped of its shell for
ornaments. Yet they must run these risks, for their eggs
would not hatch without the warm sun, and we see how
great is the gap between the last water-breathers and the
first air-breathers, when we remember that the frogs go
back to lay their eggs in the water, while the tortoises, even
when they live far out at sea, are forced to come in to shore,
in spite of great dangers, to lay their eggs that their little
ones may begin life upon land.

* * * * *

And now, if we leave the tortoises and turn to the Lizards,
we find them meeting life's difficulties in quite a different
way. Here are no sluggish movements, horny beaks, and

78   Chelonia midas.
79   Chelonia imbricata.

Fig. 23.

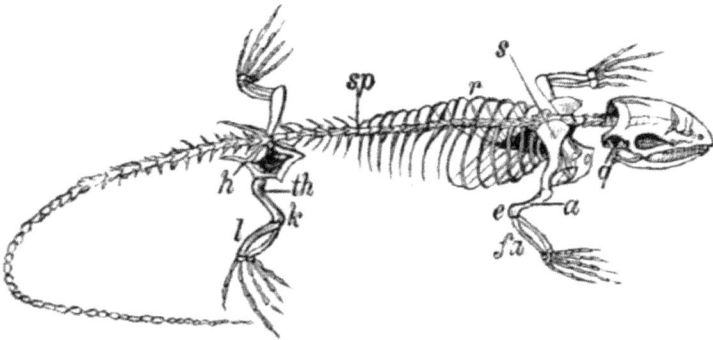

Skeleton of a Lizard.

sp, Spinous processes, which in the tortoise are flattened into plates; *r*, ribs; *s*, shoulder bone; *a*, upper arm; *e*, elbow; *fa*, forearm; *h*, hip bone; *th*, thigh bone; *k*, knee; *l*, bones of the leg; *q*, quadrate bone between upper and lower jaw.

strong boxes; but bright-eyed creatures covered with shining scales, their mouths filled with sharp teeth, with which even the small lizards can bite fiercely, and having nimble lissome bodies, which wriggle through the grass or up the trees in the twinkling of an eye. Yet the lizards, as we have seen, are formed on the same plan as the tortoise, and their scales are thickenings in their outer skin, just as his tortoise-shell is, and not true scales like those of fish. They have learned to hold their own by sharpness and quickness, and are probably the most intelligent of all the cold-blooded animals, though even they are only lively in a jerky way under the influence of warmth. They can breathe more easily than the tortoise, for their ribs rise and fall, drawing in and driving out the air they need; but they are still cold-blooded, for their heart has only three chambers. It is when the bright sun is shining that they love to dart about, chasing the insects upon which they feed; and the joints of their backbone move so easily upon

Fig. 24.

Gecko and Chamæleon.

each other that they can twist and turn in all imaginable ways, keeping their heads twisted in a most comical manner when on the watch for flies. Nay, the very vertebræ themselves are so loosely made that they can split in half, and if you seize a lizard by the tail he will most likely leave it in your hand and grow another.

They can live both in dry sandy places, where larger animals cannot find food and water, and in thick underwood, and marshy unhealthy places, where more quickly-breathing animals would be poisoned by the fetid air; and we find them swarming in hot countries in spite of enemies, their scales protecting them from the rough surface of the rocks and trees on which they glide, their feeble legs scarcely ever lifting their body from the object on which they glide rather than walk.

The true land-creepers, like our little Scaly Lizard,[80] lurk in dry woody places, and on heaths and banks, darting out on the unwary insects. Many of them lay their eggs in the warm sand or earth, but the Scaly lizard carries them till they are ready to break, so that the young ones come out lively and active as the eggs are laid. Others have taken to the water, and among these are the Monitors of Africa and Australia, which feed on frogs and fish and crocodiles' eggs, and are so strong and fierce that they often drag larger animals under the water. Some are tree and wall climbers, such as the "Geckos," with thick tongues and dull mottled skins, and they have sharp claws and suckers under their toes, so that they can hang or walk upside down, on ceilings or overhanging rocks, or on the smooth trunks of trees; and they love to chase the insects in the hot sultry nights, tracking them to their secret haunts. They are far more active than the large gentle Iguanas or Tree-Lizards of South America, from a few inches to five feet long, which may be seen among the branches of the trees of Mexico, their beautiful scales glistening in the sun as they feed on the flowers and fruit. They swarm on all sides in those rich forest regions, scampering over the ground, and then clinging with their claws to the tree-bark as they gradually mount up into the dense foliage; and they have many advantages, for not only can they climb to great heights out of the reach of beasts of prey, but they can also swim well, having been known to fling themselves from the overhanging branches into the water below when danger was near. They do not, moreover, descend as gracefully as the "Flying Lizards" of the East Indies, which have a fold of skin stretched from the lengthened ends of their hinder ribs, so that they sail from branch to branch as they chase the butterflies and other insects.

But the most curious of all tree-lizards is the Chamæleon,

with his soft warty skin, his round skin-encircled eyes, his
bird-like feet, and his clinging tail. He never hurries himself, but putting forward a leg, at the end of which is a foot
whose claws are divided into two bundles, he very deliberately grasps the branch, as a parrot does, loosens his tail,
draws himself forward, and then fastens on again with tail
and claws; while his eyes, each peering out of a thick covering skin, roll round quite independently of each other, one
looking steadily to the right, while the other may be making
a journey to the left. What is he looking for? Just ahead
of him on a twig sits a fly, but he cannot reach him yet. So
once more a leg comes out, and his body is drawn gradually
forwards. Snap! In a moment his mouth has opened, his
tube-like tongue, with clubbed and sticky tip, has darted out
and struck the fly, and carried it down his throat, while the
chamæleon looks as if he had never moved. It is not difficult
to imagine that such a slow-moving animal, whose natural
colour is a brownish green like the leaves among which he
moves, would often escape unseen from his enemies. And
when light falls upon him, his tint changes by the movement
of the colour-cells in his skin, which seem to vary according
to the colour of the objects around, whenever he is awake
and can see them.

So by the waterside, on the land, and among the trees, the
lizard tribe still flourish in spite of higher animals; and just as
we found some legless kinds among the *amphibia* burrowing
in the ground, so here, too, we find legless lizards, some with
small scaly spikes in the place of hind legs, others, like the
glass-snake of America[81] and our English slowworm[82] (or
blindworm), which have no trace of feet outside the skin, but

81   Ophisaurus ventralis.
82   Anguis fragilis.

glide along under grass and leaves, eating slugs and other small creatures, though they are true lizards with shoulder bones and breastbones under the skin.

* * * * *

Here, then, we seem to be drifting along the road to snake-life, but we must halt and travel first in another direction, upwards to a higher group of animals, which may almost be called gigantic flesh-eating lizards, though they are far more formidable and highly-organised creatures. These are the Crocodiles, and no one looking at them can doubt for a moment that they at least are well armed, so as to have an easy time of it without much exertion. Huge creatures, often more than twenty feet long, with enormous heads

Fig. 25.

The Nile Crocodile.—(*Tristram.*)

and wide-opening mouths, holding more than thirty teeth in each jaw, they look formidable indeed as they drag their heavy bodies along the muddy banks of the Nile, their legs not being strong enough to lift them from the ground. Their whole body is covered with strong horny shields, and under these shields, on the back, are thick bony plates, which will turn even a bullet aside, and quite protect the crocodile from the fangs of wild beasts. Their eyelids are thick and strong, and they have a third skin which they can draw over the eye sideways like birds; their ears, too, have flaps to cover them, and their teeth are stronger and more perfect than any we have yet seen, for they are set in sockets, and new ones grow up inside the lower part of the old ones as they are broken or worn away.

But it is in the water that we see them in their full strength; there they swim with their webbed feet and strokes of their powerful tail, and feed upon the fishes and water animals—monarchs of all they survey. Nor is the crocodile content with mere fish-diet. Often he will lie with his nostrils just above the water and wait till some animal—it may be a goat, or a hog, or even a good-sized calf—comes to drink, then he will come up slowly towards it, seize it in his formidable jaws, or sometimes strike it with his powerful tail, and drag it under water to drown. For he himself can shut down his eyelids and the flaps over his ears, and he has a valve in the back of his throat which he can close, and prevent the water rushing down his open mouth; and after a while he rises slowly till his nostrils are just above the water, and he can breathe freely while his victim is drowning, because his nose-holes are very far back behind the valve. Then when it is dead he brings it to shore to tear it to pieces and eat it.

Thus the crocodiles of the Nile and the Ganges, the

Gavials with their long narrow snouts, and the Alligators of America, with their shorter and broader heads, feed on fish and beasts, and all dead and putrid matter, acting as scavengers of the rivers; while they themselves are almost free from attack, except when tigers fall upon them on land. But it is the young crocodiles which run the most risks when they come out of the small chalky eggs which have been hatched in the warm sand of the shore. True, their mother often watches over them at this time, and even feeds them from her own mouth; but in spite of her care many of them are eaten in their youth by the tortoises and fishes which they would themselves have devoured by-and-by, if they had lived to grow up; while the monitors, ichneumons, waterfowl, and even monkeys, devour large numbers of crocodiles' eggs.

* * * * *

And now, if we were to turn our backs upon the great rivers in which these animals dwell, and wander into the Indian jungle or the South American forest, we might meet with enemies far more dangerous and deadly, although they stand much lower in the reptile world. Who would think that the huge boa of South America, and the python and poisonous cobra of India, or even our own little viper, whose bite is often death to its victim, are creatures of lower structure than the harmless little lizard or the stupid alligator? Yet so it is. For Snakes have no breastbone and have lost all vestiges of front legs and shoulder bones, nor have they any hips or hind legs except among the boas and rock-snakes; and even these have only small traces of hips, which carry some crooked bones, ending in horny or fleshy claws, in the place where hind legs ought to be. They have no eyelids (and by this we may know them from the legless lizards), but their skin grows right over the eyes, so that when a snake

Fig. 26.

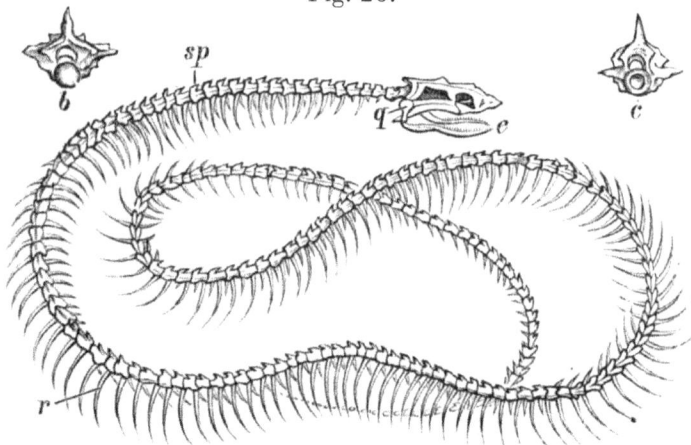

Skeleton of a Snake.

sp, Spinous processes of the joints; *r*, ribs; *q*, quadrate bones, joining upper
and lower jaws; *e*, front of the lower jaw, where there is an elastic band
in the place of bone; *b*, ball end of joint, facing the tail; *c*, cup end of joint,
facing the head.

casts its skin there are no holes where the eyes have been,
but only clear round spaces like watch-glasses, in the scaly
skin. Their ears have no drum, and are quite hidden under
the scales with which their body is so thickly covered that
they must feel very little as they glide along. These scales,
like those of the lizard, are thickened parts of the outer skin,
and if you stretch a piece of snake-skin you can see them
lying embedded in it, the clear skin itself showing between.

We must not, however, imagine that the snake is at a
disadvantage because he has lost so many parts which other
reptiles possess. On the contrary, he has most probably lost
them because he can do better without them. The transparent tough skin over his eye is a far better protection in narrow rugged places, and among brakes and brambles, than
a soft movable eyelid; and if he does not see as well as the
crocodile, he has a most delicate organ of touch in his long,

narrow, forked tongue, with which he is constantly feeling as he goes, touching now on one side, now on the other, each object he comes near, and drawing the tongue in at every moment to moisten it in a sheath at the back of his throat. A breast bone, moreover, would have been a decided hindrance to him, for he wants the free use of all his ribs; and as to the loss of his legs—in the place of four he has often more than two hundred. For all along his backbone, except just at the head and tail, a pair of ribs grow from each vertebra, being joined to it by a cup-and-ball joint (*c* and *b*, Fig. 26), and the muscles between them are so elastic that the ribs can be drawn out so that the body seems to swell, and then drawn back towards the tail. In doing this they strike the ground and the snake moves forwards, just as a centipede does on its hundred legs.

It is worth while to take our harmless Ringed Snake in your hand to feel this curious movement to and fro of the ribs, and to notice how the creature forces itself through your grasp. Moreover, you will learn at the same time one use of the broad single plates under the snake's body (see Fig. 27), for they, like all the scales, are loose from the skin on the side towards the tail; and as they are fastened by muscles to the ends of the ribs, you will find that at each movement they stand up a little like tiles on a roof, and their edges coming against your hand help to drive the snake forward.

Another thing you will learn if the snake does not know you, and that is how strangely they hiss, often with their mouth closed, while their whole body seems to quiver. This is very puzzling at first, till you learn that one of their lungs has shrunk up, and the other is a very long and narrow bag stretching nearly the whole length of the snake's stomach, and the hissing sound is made by drawing in and forcing out the air from this long bag.

Fig. 27.

COMMON RINGED SNAKE.[83]
Where the body is coiled the single under plates are seen.

Meanwhile, another way in which the snake will escape from your hold unless you grasp it tightly, is by wriggling in all directions, so that you do not know where to expect it next; for the whole of the joints of its backbone are joined by a succession of cups-and-balls, the ball of one joint fitting into the cup in the one behind it. It is easy to see how such joints can move almost every way, since the ball can twist

83   Natrix torquata.

freely in the cup wherever the muscles pull it (except where checked by the spines on the top of the backbone), and can even turn so much to one side that the snake can coil itself round or tie itself into a knot.

A creature that can glide along so smoothly, twist about so freely round trees, through narrow openings and tangled brushwood, and even swim in the water, has no small advantage in life; and the snake can also coil itself up under a heap of dead leaves or in a hollow trunk of a tree for safety, or to watch for its prey when no animal would suspect it was near. But even the harmless snakes have something besides this, namely, the power of swallowing animals much broader and thicker than themselves. You will see on looking at the lizard's skull (p. 103) that its bottom jaw is not joined at once to the top one, but there is a bone (q) between, which enables it to open its mouth wider than if the two jaws touched each other. Now this bone (q) in the snake's jaw is so loosely hung that it moves very easily, and the lower jaw also stretches back far behind the upper one, so that when the snake brings the jaw forward it can open its mouth enormously wide. Nor is this all; it can actually stretch the bones of its jaws apart, for they have not their pieces all firmly fixed together. In the front of the mouth each jaw has elastic gristle in the place of bone, and the two halves of the jaw can thus be forced apart from each other, making room for a very large mouthful indeed.

Now the snake's teeth are all curved towards the back of his mouth, and they are never used for chewing or tearing, but only for holding and packing down its food. So when he seizes a creature too large to be easily swallowed, he fastens his front teeth into it and then brings forward *one* side of his jaws. He then fixes the teeth of this side into the animal, and

Arabella B Buckley

Fig. 28.

The Boa Constrictor in the Forests of South America.

holds it fast while he brings forward the jaws on the *other* side, fixes these teeth, then loosens and brings forward the others, and so on. In this way he keeps his mouth stretched over the prey and gradually forces it down his elastic throat, moistening it well all the time with slime from two glands, one on each side of his mouth, and when it is swallowed he lies down and rests while the stomach digests its heavy load.

We see, then, that even harmless snakes have many advantages. Thus our ringed snake, feeding on mice and lizards, frogs and fish, wanders through the grass and bushes of warm sunny banks, feeling this side and that with his delicate forked tongue, and gliding so fast that the lizards and mice try in vain to escape; while in the water he seizes the frogs by their hind legs and jerks them into his mouth. He does not even always stop to kill his food, for a live frog has been known to jump out of a snake's mouth as it yawned after its meal. So he lives through the summer, changing his skin several times by loosening it first at the lips, so that two flaps lie back over the head and neck, and then rubbing himself through moss, bush, or bramble, so that the skin is drawn off inside out like a glove, and the new skin appears underneath, fresh, hard, and bright, ready for use. Then in the warm season the mother lays her ten or twenty soft eggs in a mass of slime, and leaves them in some sunny spot, or under a heap of warm manure to hatch, and she herself wanders away, and when winter comes coils herself up in the trunk of some hollow tree, or under the hedge, to sleep till spring comes round again. Life does not always, however, flow so smoothly as this, for the snakes have their enemies; the fox and the hedgehog love to feed upon them, the buzzard and other birds of prey swoop down upon them from above, and the weasels attack them below; and this, perhaps, is partly the reason why the ringed snake generally keeps near the water, into which it can glide when danger threatens.

All snakes are not, however, so harmless as our little

ringed snake. The Pythons of India and the Boas of America, though they have no poison in their teeth, can work terrible mischief with their powerful joints as they coil round even good-sized animals, such as an antelope or a wild boar, and crush them in their folds. Then it may be seen what a terrible weapon this flexible backbone is, as the muscles draw it tighter and tighter round the unfortunate animal, breaking its bones in pieces, till, when it is soft enough to be swallowed, the snake gradually forces it down its capacious mouth, moistening it with saliva as it goes. These large boas and pythons would, in fact, probably devastate whole countries if it were not that when they are young they are devoured by other animals, so that very few live to grow into dangerous marauders.

Other snakes have taken a still more terrible way of killing their prey. There may be some chance of escape from a coiling snake, unless he already holds you with his teeth, but the poisonous Cobra[84] may strike before you know that you have startled him, and though the Rattlesnake[85] makes a sharp noise as he shakes the loose horny plates to call his mate or to alarm an enemy, yet when he means to strike his prey it is too late when the sound is heard to get out of reach of his fatal fangs. From the snake's point of view, however, it is clearly an advantage to be able with one single stroke to paralyse its prey, so that it has only to wait for the poison to do its work, and then its meal is ready. Even our little viper (see p. 120), needs only to strike a mouse once, and then draws back as the poor victim springs up and falls and dies, soon to be packed down its destroyer's throat.

84   Naja.
85   Crotalus.

Fig. 29.

The Cobra di Capello.[86]—(*FROM GOSSE.*)

The mouth being closed, the poison fangs cannot be seen. The tongue is
perfectly harmless.

Yet this terrible poison, which acts so speedily, is no
special gift to the snake. It has only lately been discovered
by M. Gautier that we, and probably all animals, have in
our saliva some of the very poison with which the cobra
kills its prey, only with us it is extremely diluted, and is
useful in digesting our food. The cobra, however, has the
poison, which no doubt exists in the slimy saliva of all
snakes, specially concentrated and collected in two glands,
one on each side of its jaw. From each of these glands
(*g*) a small canal passes under the eye to the edge of the
jaw (*c*), and opens immediately above a large curved fang

86   Naja tripudians.

Fig. 30.

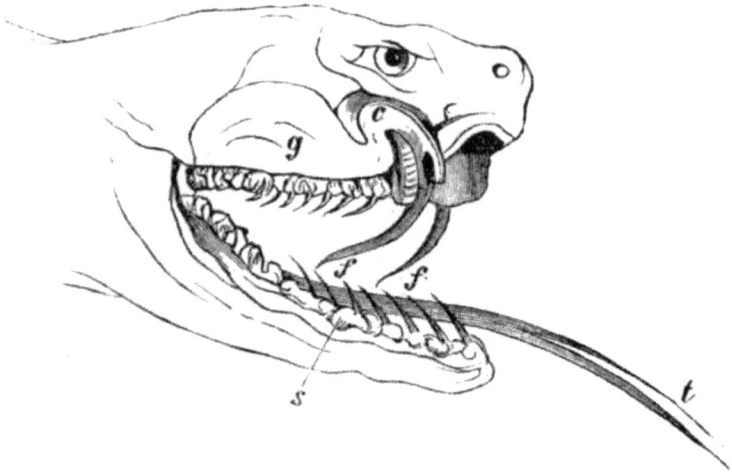

Jaw of a Rattlesnake.
ff. Poison fangs; *g*, gland secreting poison; *c*, canal leading from gland to
base of fang; *t*, harmless tongue; *s*, saliva glands.

(*f*). This fang is fastened to a bone in the cheek which
moves easily, so that the poison teeth can be shut back
and lie close against the gum when they are not wanted,
and when they are wanted can be brought quickly down
again. Though the fang looks round like ordinary teeth,
it is really flattened out like a knife-blade, and then the
edges are curved forwards so as to form a groove or, in
some snakes, a closed tube, down which the poison can
run to the point.

Now when the snake wishes to strike its prey it raises
its head, brings down the fangs and drives them into the
creature's flesh, and at the same time certain muscles
press upon the poison gland, so that the liquid poison is
forced into the wounds. If, however, the fang was fixed
to the canal, the snake's weapon would be gone if the
point were broken, so we find that the canal-opening lies

just *above* the tube of the tooth, and behind are six small reserve teeth, covered by a tender sheath skin, ready to grow up and take its place when wanted.

Should we not think that with such weapons as these the poisonous snakes would conquer every enemy? Yet they, too, only have their fair chance of life, for besides the destruction of their eggs other dangers await them. The rapacious birds, with their feathery covering, their horny and scale-covered legs and feet, and their hard beaks, will offer battle even to a poisonous snake. The buzzard makes short work of our common viper or adder, whose fangs, though fatal to small animals, are not nearly so powerful as those of snakes of hot countries. Seizing the viper with his claws in the middle of its body, the buzzard takes no notice of its frantic struggles as, winding itself about his feet, and striking wildly at his breast, his wings, and his scaly legs, it

> "... doth ever seek
> Upon its enemy's heart a mortal wound to wreak."

Keeping his own head well back out of danger, the bird lets the snake exhaust itself, waiting only till he can give a fatal blow with his beak upon its upraised head, and then, soon despatching it, tears it to pieces for a meal. Nor is even the dreaded Cobra safe from danger, for he finds his match in the powerful Adjutant birds (see p. 127), and in the Indian Ichneumon or Mungoos, which attacks the snake boldly, skilfully dodging the fatal stroke until it has broken the neck of its enemy; while in Africa the bold Secretary bird is complete master of the dreaded poisonous snakes of that country. In fact, there is little doubt that every kind of snake, either in youth or age, falls a victim to some kind

Fig. 31.

Common English viper (*Pelias berus*), with poison-fangs showing in the
open mouth, and the soft harmless tongue outstretched to feel.

of bird or beast; and even the poisonous sea-snakes, which
swarm in the tropical seas, probably find their masters in
the pugnacious saw-fish and the thick-skinned shark.

We see, then, that it is not without some struggle that
these cold-blooded reptiles have held their own in the world,
nor is it to be wondered at that only these four types—tortoises, lizards, crocodiles, and snakes—should have managed
to find room to live among the myriads of warm-blooded
animals which have filled the earth. These four groups have
made a good fight of it, and many of them even make use
of warm-blooded animals as food. The tortoises, it is true,
feed upon plants, except those that live in the fresh water,

and feed chiefly on fish, snakes, and frogs, while most of the lizards are insect-feeders. But the crocodile, as he lurks near the river's edge, and the snake, when he fastens his glittering eye on a mouse or bird, are both on the look-out for animals higher in the world than themselves.

It is, perhaps, natural that we should shrink from cold-blooded creatures, especially as they *seem* to show very little affection. Yet lizards, tortoises, and snakes can all be made to know and care for those who are kind to them; while, as we have seen, the fierce crocodile watches over and feeds her young, and the python coils herself over her eggs, and will take no food till they are hatched. Moreover, we can scarcely look at the quaint shell-covered tortoise, or examine the heavily-mailed coat of the alligator, or the poison-fangs of the snake, without admiring the curious devices by which these animals have managed to survive in a world which once belonged to their ancestors, before our present swarm of warm-blooded animals multiplied and took possession of their kingdom.

THE EARLIEST KNOWN WATERBIRDS

CHAPTER VI.

THE FEATHERED CONQUERORS
OF THE AIR.

Part I.—Their Wanderings over Sea
and Marsh, Desert and Plain.

IT is a warm sunny day in early spring, one of those few
bright days which sometimes burst upon us in April, just

after the swallows have come back to us, searching out their old nooks under the eaves, or their old corners in the chimneys, to build their new nests. There they are, clinging with their sharp claws to the edge of the cottage thatch, while the impudent little sparrow, which has remained hopping about all the winter long, chirrups at them from a neighbouring apple-tree. Upon the grass-plot near a blackbird is pecking at a worm, and from the wood beyond a thrush trills out his clear and mellow song, accompanied from time to time by the distant cry of the cuckoo calling to his mate. For it is the love-time of the birds; and as we watch them flying merrily hither and thither in the bright sunshine, we ask ourselves whether we must not have made a great leap on leaving the cold-blooded snakes and tortoises, since now we find ourselves among such merry, warm-hearted, passionate little beings, with their beautiful feathery plumage, their light rapid flight, their love for each other, their skill in nest-building, and their patient care for their little ones.

And, indeed, we have come into quite a new life, for now we are going to wander among the conquerors of the air, who have learned to rise far beyond our solid ground, and to soar, like the lark, into the clouds, or, like the eagle, to sail over the topmost crags of the mountains, there to build his solitary eyrie.

Even the little sparrow, which flits about by the roadside, can laugh at us with his impudent little chirp, as he flies up out of reach to the topmost branch of a tree. And yet a glance at his skeleton will show us that he has the same framework as a reptile, only it is altered to suit his mode of life.

True, his breastbone (*b*, Fig. 33) is deep and thin instead of flat, and those joints of his backbone which lie between his neck and tail are soldered firmly together, more like those of

Fig. 32.

THE SPARROW.

With wings raised, as in the skeleton on next page.

the tortoise, and he stands only upon two feet. Yet this last difference is merely apparent, for if you look at the bones of his wings you will find that they are, bone for bone, the same as those in the front legs of a lizard, only they have been drawn backwards and upwards so as to work in the air.

There is the upper arm (*a*) answering to the same part of the lizard's front limb (p. 103); there is the elbow (*e*); then

Fig. 33.

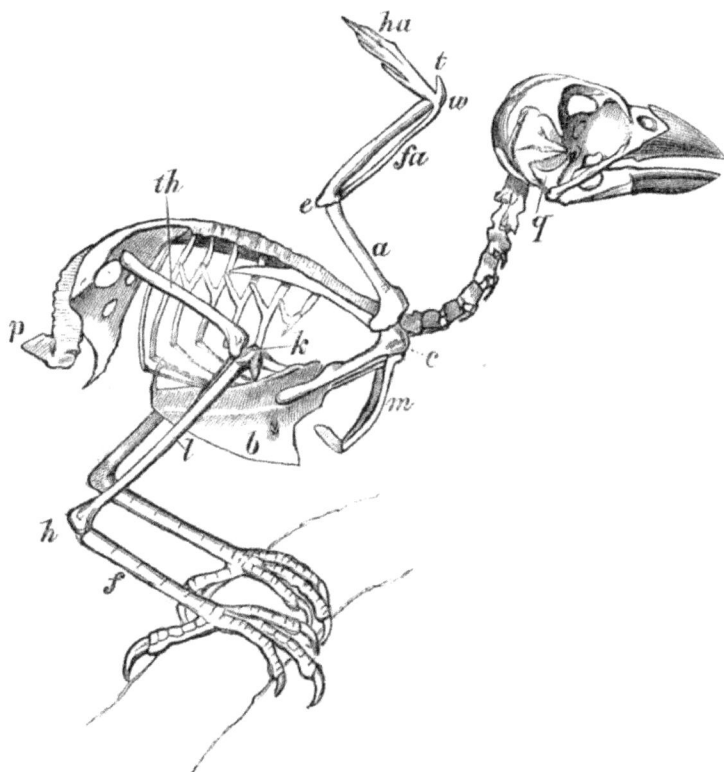

Skeleton of a Sparrow (from a specimen).

q, Quadrate bone, peculiar to reptiles and birds and some *amphibia*; *b*, breast-bone; *m*, merrythought or collar bone; *c*, coracoid bone, over which the tendon works to pull up the wing; *p*, ploughshare bone, on which the tail grows. Wing bones—*a*, upper arm; *e*, elbow; *fa*, fore arm; *w*, wrist; *t*, thumb; *ha*, hand. Leg bones—*th*, thigh bone; *k*, knee; *l*, lower part of leg; *h*, heel; *f*, foot.

the two bones of the fore-arm (*fa*); then the wrist (*w*), and a long hand (*h*), which has lost almost all trace of separate fingers, except the little thumb (*t*), which carries some feathers of its own, known as the "bastard" wing. Now when the sparrow is resting he draws back his elbow, folds his wrist joint, and brings the whole wing flat to his body. But when he wishes to fly he stretches his arms out and beats the air with them, and as his hand moves over most space, it is there that you will find the longest quill feathers, which stretch right to the tip of his wing; then next to these follow the feathers of his fore-arm, while those of the upper arm are short and close to his body, and over all these are the rows of covering feathers, which make the whole wing thick and compact.

Here, then, we have the lizard's front legs turned into a wonderful flying machine in the bird, and this in *quite a different way* from the flying lizards which lived long ago, and which had only a piece of membrane to flit with, like bats. And now what has happened to the hind legs, the only ones used as legs by the birds? Look at the sparrow as he clasps the bough with his toes, and you will, perhaps, be puzzled why the first joint of his leg turns back like an elbow and not forward like a knee. Ah! but that joint is his ankle, and the knob behind is his heel (*h*), for the bones of his foot have grown long and leg-like; and he always stands upon his toes, the rest of his foot forming a firm support to hold his body up in the air. Look at the skeleton and you will find his true knee (*k*) up above; and if you go to the Zoological Gardens and watch the Adjutant birds, you will often find them resting their whole foot upon the ground (see Fig. 34), and comical as it looks, it will help to explain the curious foot and leg of a bird.

Already, then, we see that the bird is using the same bones as a reptile, though he uses them in a different way; and besides these resemblances between the skeletons of birds and reptiles there are two special ones easy enough for us to understand. We saw in the snakes and the lizards that they have a separate bone (9, Figs. 23 and 26) joining the lower jaw on to the head; now you will find this same

Fig. 34.

The Adjutant Bird.

Showing the foot resting from heel to toe upon the ground.

bone in the sparrow and in all birds (see Fig. 33), but in quadrupeds this bone is not to be found, the part representing it being changed into one of the bones of the ear. Again, the sparrow's skull is joined to his backbone by a single half-moon-shaped knob, which fits into a groove in the first joint or vertebra. This also we find in reptiles, while all higher animals have two such knobs, so that although they can nod the head upon these, they cannot turn it upon them, and consequently the first joint turns with the skull upon the second vertebra.

These, then, are some of the reasons why Professor Huxley tells us that though frogs and reptiles look in many ways so like each other, yet in truth the frogs must be grouped with the gill-breathing and fish-like animals;[87] while the cold-blooded reptiles, when we come to look closely into them, are linked with such different looking creatures as the bright and merry birds.[88] But we have also another and stronger reason for thinking that reptiles and birds are distant connections; for in those far bygone times (see p. 89), when the huge land-lizards browsed upon the trees, the birds living among them were much more like them in many ways than they are now. From their skeletons and feathers which we find, we know that the strange land birds[89] which then perched on the trees had not a fan-shaped tail made of feathers, growing on one broad bone as our birds have now (*p*, Fig. 33), but they had *a long tail of many joints like lizards*, only that each joint carried a pair of feathers, and like lizards too they had *teeth in their jaws*, which no living bird has. They must have been poor flyers at best, these earliest known birds, for their wings were small and the fingers of their hand were separate more like lizard's toes, two of them at least having claws upon them, while their long hanging tail must have been very awkward compared to the fan-shaped tail they now wear. For some time they were the only birds we know of, but later on we come upon the bones of water-birds[90] telling the same story. For

87   Ichthyopsida—*ichthys*, fish; *opsis*, appearance.
88   *Sauropsida—sauros*, lizard; *opsis*, appearance.
89   Archœopteryx, see picture-heading, Chapter VII.
90   See picture-heading of this chapter.

some about the size of small gulls,[91] though they flew with strong wings and had fan-shaped tails, still had teeth in their horny jaws, set in sockets like those of the crocodile, while their backbones had joints like those of fishes rather than birds; and with them were other and wingless birds[92] rather larger than our swans, but more like swimming fish-eating ostriches, for their breastbones were flat, not thin and sharp like the sparrow's, and they had scarcely any wings, short tails, long slender necks, and jaws full of teeth, this time set in grooves like those of lizards and snakes.

In these and many other points the early birds came very near to the reptiles—not to the flying ones, but to those which walked on the land. And now, perhaps, you will ask, did reptiles then turn into birds? No, since they were both living at the same time, and those reptiles which flew did so like bats, and not in any way like the birds which were their companions. To explain the facts we must go much farther back than this. If any one were to ask us whether the Australian colonists came from the white Americans or the Americans from the Australians, we should answer, "neither the one nor the other, and yet they are related, for both have sprung from the English race." In the same way, when we see how like the ancient birds and reptiles were to each other, so that it is very difficult to say which were bird-like reptiles and which were reptile-like birds, we can only conclude that they, too, once branched off from some older race which had that bone between the jaws, that single neck joint, and the other characters which birds and reptiles have in common.

But long ago they must have gone off each on their own

91   *Ichthyornis*, fish-bird.
92   Hesperornis.

road, the reptiles filling the world for a time, flying and walking and swimming, till they found at last that creeping was their most successful way of life; the birds on the other hand becoming more and more masters of the air and the water, so that, while keeping the same bones and parts as the reptiles, they have grown into quite different beings in their form and habits, giving up the long-jointed tail of the *Archœopteryx*, or ancient-winged bird, for the compact feathered fan which helps to balance them in their flight, and the teeth of the water-birds for the sharp and horny beak, which, together with their claws, is their chief weapon of attack and defence now that they have employed their front limbs as wings.

Nor shall we have far to look for the secret of their success in life. Just as the reptiles have an advantage over the naked frogs and newts by having strong scaly coverings in their skin, so the birds have an advantage over the reptiles in that beautiful feathery plumage which covers their body, and the powerful muscles which work their limbs. For it is by means of these that they have been able to move quickly and travel far, and to develop that bright nervous intelligence which has grown more and more active as they have been carried into fresh scenes and experiences, overcoming new difficulties and enjoying new pleasures.

Remember for a moment how weak the lizard's limbs are, so that his body always drags upon the ground; and then look at the bird's tight grasp of the bough and the rod-like legs which raise his body above it. Watch him as he beats the air with his wings, rising and sinking, turning and swerving at will, and you will see that he has earned freedom, strength, and active life, by means of the strong muscles which move these legs and wings, and the feathers which provide him with an instrument for beating the air. Feel a sparrow's fat little breast, or see how much meat comes off the wing and breast of a pigeon, and then, if you consider that all this flesh is muscle used for moving his wings, you will not wonder at his easy flight. For the muscles of a bird's breast often weigh more than all his other muscles put together, and while one enormous mass of muscle in front of the breast works to pull down the wing, another smaller one, ending in a cord or *tendon*, passing like a pulley over the top of a bone (*c*, Fig. 33, p. 125), pulls it up, so that by using these, one after the other, the bird flies.

But where have the feathers come from,—those wonderful beautiful appendages, without which he could not fly? They are growths of the bird's skin, of the same nature as the scales of reptiles, or those on the bird's own feet and legs; and on some low birds such as the penguins they are so stiff and scale-like that it is often difficult to say where the scales end and the feathers begin. All feathers, even the most delicate, are made of horny matter, though it splits up into so many shreds as it grows that they look like the finest hair, and Dr. Gadow has reckoned that there must be fifty-four million branches and threads upon one good-sized eagle's feather.

When these feathers first begin to grow they are like little grooved pimples upon the flesh, then soon these pimples sink in till a kind of cup is formed all round them, and into this cup the soft layer just under the outer skin sends out fibres, which afterwards form the pith. Round these fibres rings of horny matter form, and then within these rings, in the grooves of the soft pimple, the true feather is fashioned. First the tips of the feathery barbs, then the shaft, and then the quill appear, as the feather grows from below, fed by an artery running up into the pimple; till at last, when the whole is full-grown, the quill is drawn in at the base, and rests in its socket, complete.

Some of these feathers are weak and soft, with slender shafts and loose threads growing all round them, and these are the downy feathers which lie close to the body and keep the bird warm. Others, which cover the outside and form the wings and tail are flat, with strong quills and shafts, and a double set of barbs growing upon each shaft; and if you look at these wing feathers under a strong microscope you will see that they have a special arrangement for helping them to resist the air. For not only have all the little featherlets or *barbs* rows of other featherlets or *barbules* growing upon them, but these again are covered with fine horny threads, often hooked at the tip, which cling to the next barb, so that the feather is woven together as it were, in a close web, and if you strike it against the air you will find that it resists it strongly.

Now in a bird's wing the feathers are so arranged that they lap one under the other from the outside of the wing to the body, so that when the bird strikes downwards they are firmly pressed together, and the whole wing, which is hollow like the bowl of a spoon, encloses a wingful of air, and as this is forced out behind, where the tips of the feathers are yielding and elastic, he is driven upwards and forwards. When, however, he lifts his wing again, the feathers turn edgeways and are separated, so that the air passes through them, and he still rises while preparing for the next stroke. All this goes on so rapidly that even the heron makes 300 strokes in a minute, and the wild duck 500, while in most birds they are so rapid that it is impossible to count them; yet all the while the little creature can direct his flight where

he will, can pause and direct his wings to the breeze so as to
soar, can swoop or hover, wheel or strike, guiding himself by
the outspread tail and a thousand delicate turns of the wing.

All this complicated machinery, however, would not have
served the bird much if his body had been as heavy, and his
blood as cold, as those of the lizard and the crocodile. But
here he has made a great step forward. In the first place,
he has a heart with four chambers, two on the right side
and two on the left; and while one of those on the right side
receives the worn-out blood from the body and pumps it *to*
the lungs to be refreshed, one of those on the left side receives
it *from* the lungs when it is refreshed, and the other pumps
it into the arteries to feed the body. So here we see for the
first time among our backboned animals a creature whose
good and bad blood are never mixed in the heart (compare
pp. 26 and 76), but it gets all the benefit possible from its
breathing, and the blood is kept fresh and pure.

Moreover, a bird's lungs are large, and are continued
into several large air-sacs, which in their turn open out
into tubes which carry air actually *into the bones*, many of
which are hollow instead of containing marrow like those
of other animals.[93]

And now we begin to see how wonderfully these little
creatures are fitted for flying. With all this air within them,
not only is their blood kept hot by constant purifying, but
their bodies are much lighter than if their bones were solid,
and they can present a much broader surface to float upon
the air without increasing equally in weight. Meanwhile, their
feathery covering prevents the cold air around from chilling

93   Some chamæleons and geckos also have air-tubes passing from the lungs into
the body, and the crocodile's skull is full of air-cells; but the two phenomena are not connected as in birds, and other parts of the skeleton or of the skin-covering, being heavy,
have a counteracting effect.

them, so that they are not only warm-blooded animals, but actually warmer-blooded even than ourselves.

Thus, then, Life has spread her feathered favourites over the world. For them there are no limits except the extreme depths of the water below, and the height beyond the atmosphere above. Wherever air-breathing creatures can go, there some bird may be found. On the dizzy ledges of inaccessible cliffs, on the wide bosom of the open ocean, on the sandy wastes of the desert, in the tops of the highest trees, on the cloud-capped peaks of the mountains, diving or swimming, flying or soaring, running, perching, darting, or sailing for miles and miles without one moment's rest, they find their way everywhere, and there is no spot from the icebound countries of the Arctic zone to the warm bright forests of the tropics where they do not penetrate; while their sharp eyes, kept free from dust and harm by a third eyelid moving rapidly sideways,[94] see far into the distance, and thus as they soar into the sky they have a power, possessed by no other animals, of overlooking a wide domain. Nor have they been obliged, like the reptiles, to take up strangely different forms to suit their various habits, for so wonderfully does their body meet all their wants that very slight changes, such as a broad body and webbed feet for the swimmers, long bare legs for the waders, a long hind toe for grasping in the perchers, and sharp claws and beak for the birds of prey, fit each one for his work, and are some of the chief distinctions we can find between them.

Even the heavy running birds, the Ostriches of Africa, the Rheas of South America, and the Emus and Cassowaries of Australia, still remain truly bird-like, though their wings

---

94   This third eyelid is a fold on the inner side of the eye; some reptiles and amphibians have it, and so have the marsupials and many of the higher animals.

are unfit for flight. True, their breastbones are flat instead
of keel-shaped, for they have no need of strong muscles to
move their wings, which now serve only as sails to catch the
wind as they run, and in many other ways they are an older
type than our flying birds; but their wing bones are formed
as if they were used for flying, and their feathers, though
loose and downy because they have no little booklets, are
like those of other birds.

Fig. 35.

The Ostrich[95] at full speed.

Strong powerful creatures they are, even in confinement.
Yet how little can we picture to ourselves, when we see the
Ostrich trotting round his paddock in the Zoological Gardens,

95   Struthio camelus.

with his wings outspread, what he is when he courses over
the free desert!—

> "Where the zebra wantonly tosses his mane,
> With wild hoof scorning the desolate plain;
> And the fleet-footed ostrich over the waste
> Speeds like a horseman who travels in haste."

There the soft pads under the *two toes* of each foot rebound
from the yielding sand as his well-bent legs straighten with
a jerk one after the other, making his body bound forward
at full speed. Then he raises his wings, sometimes on one
side, sometimes on both, to balance himself, and to serve as
sails to help him; and with this help his stride is sometimes
as great as twenty feet, and he dashes along at the rate of
twenty-six miles an hour. He is not so heavy as he looks, for
many of his bones are hollow, his feathers are downy and
soft, and his wing-bones are small; and it is to his feather-
less thighs that you must look for the strong muscles to
which he trusts for all his speed, as with outstretched neck
he bounds across the plain.

If we go back to long bygone times, before the lion, the
leopard, and other ferocious animals found their way into
Africa, we can imagine how this great running bird took
possession of the land and became too heavy for flight; while
as time rolled on, he gained that strength of body and leg
which now is his great protection as he dashes along, his
four or five wives following in his train. The ostriches can
travel over wide distances from one oasis to another, feeding
on seeds and fruit, beetles, locusts, and small animals, and
fighting fiercely with legs and beak if attacked. And when the
springtime comes the wives lay their eggs in a hole scooped
in the sand, or often in some dry patch of ground surrounded
by high grass, till sixteen or twenty are ready; and then they

take their turn (the father among the rest) of sitting upon them, at least at night, even if they leave them to the heat of the sun by day. And when six weeks have passed the father grows impatient, and, pressing the large bare pad in front of his chest against each egg in turn, breaks it, pulls out the membranous bag with the young bird in it, shakes him out, and, swallowing the bag, goes on to another. In this way the whole downy brood are soon set free, and begin picking up small stones to prepare their gizzard or muscular second stomach for grinding, while their parents scrape the sand and find and break up food for them.

So the ostrich lives its life in Africa, from Algeria right down to Cape Colony; while its smaller and lighter-coloured relations, the Rheas, with their *three-toed* feet, course over the plains of Paraguay and Brazil, on the other side of the Atlantic, often swimming from island to island, in the bays or across the rivers, but quite unable to fly with their soft hair-like feathers, though their wings are larger than those of the ostrich. Then when we turn to the East we find other running birds; the Cassowary, with its three toes, its horny helmet, its five long single feathers, and its five naked pointed quills in the place of a wing, feeding on fruit and vegetables in New Guinea, or sharing the dreary scrubs of Australia with the almost wingless Emus wandering in pairs, the only constant married couples among the running birds.

Nor is New Zealand left without a representative of this family, for there we have the curious little Apteryx or Kiwi (Fig. 36), with its thick stumpy legs, its long beak, and its soft downy body, under which are hidden its aborted wings. Perhaps it is because he is small and insignificant that the apteryx has lived on till now, crouching under the bushes by day and creeping about in the twilight, thrusting his

Fig. 36.

Wingless birds of New Zealand.
The giant Moa (*Palapteryx*) and the tiny Apteryx. The Moa is no longer to be found alive.

long nose-tipped beak into the damp ground to draw out the worms. For long ago, though in the memory of man, as we learn from the traditions of the Maories, other wingless birds called Moas,[96] which were six or seven feet high, lived in New Zealand, and had fierce fights with the natives. We find their bones now, often charred from having been cooked in the native ovens, and when they are put together they give us skeletons which show that these birds must have been as formidable as that great bird of Madagascar, the Æpyornis, whose gigantic bones and eggs, three times the size of ostrich eggs, have been found, though the bird itself has never been seen.

But these are gone now, and their relations the Emus are fast following them: for however well these flightless birds may flourish on the broad plains and deserts, where only their wild companions are around them, they are sadly

[96]   Dinornidæ, of which Dinornis, a still more ancient form, must have been ten feet high.

at the mercy of man. The proud eagle can fly far beyond
the reach of the hunter's gun; the little lark, if she be only
wary enough, may trill out her song in the blue vault above
and leave the cruel destroyer far below; but the emu and the
cassowary, the rhea and the ostrich, have lost the power to
leave the earth; and if it were not that we prize the two last
for their feathers, they, too, like their companions, might
live to rue the day when they became runners instead of
conquerors of the air.

* * * * *

It is very different, however, with the water-birds, for
they have not only kept the power of flight, but have gained
the double advantage of also floating safely on the water.
Look at our common wild duck. We might have taken him
just as well as the sparrow for our type of a bird, and yet
while the sparrow leads a land life in the trees, the duck's
home is on the water, and many of his relations live cradled
on the open ocean.

See his broad boat-like body which floats without any
effort of his; notice how closely it is covered with short
thickly-grown feathers, which protect him from the chilly
water, while he keeps them well-oiled with his beak, from
an oil-gland which all flying birds have at the base of the
tail. Watch how he swims, drawing his webbed foot together
as he brings it forward, and spreading it like a fan to strike
the water as he drives it back. Then, as he feeds, watch him
gobbling in the mud and then shaking his head as he throws
his beak up in the air. For he, like all ducks and geese, has
a set of thin horny plates inside his broad bill, and they sift
the mud, while the tender tooth-like edges of his beak and
tongue feel out the suitable morsels.

All this time he is a water animal, but when he rises

and flies he is also master of the air, for his strong wings
carry him stoutly, so that he can migrate from one pool to
another; or in winter, when the pools are frozen, to the open
sea. He is by no means the best flyer of his family, and yet
he is spread over Europe and North America, and even as
far east as Japan, while his ocean-loving cousin, the eider-
duck, lines her nest and lays her eggs high up in Arctic
latitudes, and dives and swims in the open ocean. So too
his relations, the wild swans and geese which wander in the
lakes and swamps of Lapland, Siberia, and Hudson's Bay,
feeding on water-weeds, worms, and slugs, build their nests
in the summer in the far north, and then fly thousands of
miles southwards to their winter homes, their strong wings
whirring in the air as they go.

Yet these are scarcely as true sea-birds as the divers, the
Guillemots and Puffins, the Auks and Grebes, which swim
out all round our coasts, and dive deep down to feed on the
fish below. How clumsy they are on land and how skilful in
the water! You may see numbers of guillemots and puffins
in summer on the high cliffs of the north of Scotland, or of
Puffin Island in the Menai Straits; the guillemots laying their
eggs on the bare ledges, and the puffins in holes which they
burrow in the cliffs face; and they sit so doggedly upon their
nests, and shuffle and hop along so awkwardly, that men
climbing up, or let down by ropes from above, knock them
over as they go. But wait till the eggs are hatched, and the
little ones have broken out of their shelly prison, each one
cracking his shell from inside by means of a little horny
knob, which all baby birds have for this purpose at the end
of their beak, and which falls off when they are fairly born.
Then fathers, mothers, and young ones, able to take care
of themselves as soon as hatched, launch out into the open

sea in August, and there is a sight of diving and swimming and fishing grand to behold. The awkward legs, placed so far back on their body, now serve as powerful oars and rudders to drive their smooth satiny bodies through the water. Their thin narrow legs cut through the waves like knives, while their short stumpy wings, closely laid against their down-covered bodies, keep them from being chilled, and so do the air-bubbles which are entangled in their short thick feathers, and which give their backs the appearance of being covered with quicksilver when they dive[97] after the fish below.

And then when the winter comes, those which have bred in the north fly and swim southwards to our coasts, where they are joined by the true divers and grebes which have come from the rivers and estuaries, where they have made their nests on some reedy bank or floating upon the water, and lived till their young ones are strong. This is their seafaring time; and whether near the shore, or miles out at sea, they dive and swim and make the ocean their home till spring comes round again.

Still all their roving is done chiefly by swimming, and they leave it to the Gulls and Petrels, the Terns and the powerful Cormorants and Gannets, to fly hither and thither over the wide sea. These birds have indeed reached the climax of a seafaring life, with their powerful wings, their sharp and often hooked beaks, and their short legs. They, too, feed upon the water, coming up with a fish in their mouth, but they do not dive under and swim like the guillemots. On the contrary, flying is their forte; they swoop down, and scarcely have they gone a few feet under water than they are up again, skimming on the waves as they swallow their

97   This beautiful effect may be seen from below when the guillemots are fed in any of the public aquariums.

Fig. 37.

A GROUP OF SEA-BIRDS.
1. Cormorant. 2. Black-winged Tern. 3. Gulls. 4. Puffins. 5. Guillemots.

prey, which may be anything from dead floating creatures
to living fish which have ventured too near the surface. Yet
they swim well too, and though the common gulls rarely

go more than twenty miles from the shore, they are quite at home on the open ocean, and there is no habitable part of the globe without them. Still more venturesome are the petrels:—

> "Up and down, up and down,
>   From the base of the wave to the billow's crown,
>   And amidst the flashing and feathery foam
>   The stormy petrel finds a home."

They are smaller and lighter than the common gulls, and are never so happy as when darting over the foam of an angry sea, while their more delicate relations, the Terns or sea-swallows, with their long pointed wings and forked tails, are taking shelter in the quiet bays.

Lastly, king among all sea-flying birds is the gigantic petrel, the Albatross. What a grand fellow he is when he is once on the wing, though he has some difficulty in starting. Flying onward, onward, without resting day or night, his pure white body near down to the water, his large head and short thick neck slightly bent, and his long, narrow, black wings, often measuring ten feet from tip to tip, widely outspread, he beats a few powerful strokes, and then sails along, using his head and tail as rudders to turn his wings to the wind. Often he will follow a ship for days, sailing round and round in circles, and yet keeping easily ahead, while all the time his bright eye watches the water to catch every chance of food. Jelly-fish, cuttle-fish, and real fish of all kinds, together with any dead creatures he may find afloat,—all is food to him, and his powerful beak will cut through the toughest morsel. For days and days he will fly on, never tiring, and feeding as he goes; and if he alights for a time upon the water he rises with difficulty, unless the waves are high and bear him up on their crests; and when he comes to rest it is on some island in mid-ocean, where he seeks a mate, and brings up his nestlings either on the low ground or on the top of a high mountain, in a hollow lined with grass and moss. Truly, if we look at the far-flying albatross we must acknowledge that the wings of a bird have given him the command of the sea as well as the land.

He forms a strange contrast to the curious stunted bird form which we may find in those same islands where the mother albatross lays her eggs. For there, in the islands of the South Pacific, close by the side of the albatross nest, are whole groups of strange-looking birds, the Penguins, with their fat, white, feathered breasts, their dark head and beak,

Fig. 38.

Albatrosses and Penguins.

their curious hind legs set right at the end of their body, and their small paddle-like wings, covered with short stiff feathers, quite useless for flight. We have come upon a strange story here, for our penguin is a low relation, of the guillemots and puffins whom we left in the north, and, like the great northern auk, which has now been extinct for many years, he has lost the use of his wings. He has no dangerous enemies on these rocky inaccessible islands, where he and his companions form dense penguin rookeries upon the ground, unless it be the large gulls or skuas which steal the eggs. Nor has he any need for flying, for the sea is all around him, and even if he wishes to migrate to warmer waters in winter, he does so by swimming. Therefore we find that his wings are lost to him for any flying purpose, and nothing can be more awkward than he looks, shuffling or hopping along with outstretched arms, like a fat baby, till he comes to the water's edge. But when he dives in and swims it is quite a different matter. Then his easy wavy motion, like that of a seal, shows at once that his stumpy imperfect wings are excellent fins, while his feet serve him both as oars and rudders.

Thus we have traced our swimming and web-footed birds to their extreme types—the strongest flyer in the albatross, and the lowest, most fish-like bird in the penguin; while, if we were to follow the pointed-winged frigate-bird in his flight, or see the pouched pelican in his home, we should find another group of these web-footed birds, no longer merely standing upon rocks, but perching upon the boughs of trees, and building their nests by the side of rivers in warm countries nearly all over the world, or among the mangrove bushes of the tropical islands.

* * * * *

And now, if we return to our northern shores and pause

upon the broad wet sands at low tide, we may chance to find whole flocks of active little birds hovering and running and wading in the water, or pecking on the sands; and the double-noted whistle of the Curlew, or the musical cry of the Peewit (or Lapwing), tell us at once that they are "waders,"—birds with bare legs, flat toes, and long beaks, which drop down on the muddy flats by the sea, seeking their food at the edge of the water. There they are, Curlews and Dunlins and Sandpipers, Plovers and Knots, Oystercatchers feeding on mussels and limpets, and Turnstones tilting up the lumps of mud to find food beneath. One and all they are running hither and thither, to seize here a shrimp or sandhopper or a tiny fish, there a

Fig. 39.

A group of Wading Birds.

1, Stilt; 2, Avocet; 3, Peewit; 4, Dunlins; 5, Curlew Sandpiper; 6, Sander-
ling; 7, Oystercatcher; 8, Curlew; 9, Turnstone.

worm or a sea-slug; making the most of their time before the sea comes up and covers their feeding ground.

Here we have no webbed feet or legs set far back, but three long, flat, straight toes, well fitted for walking on marshy ground and treading lightly on water-plants, and slender bodies well balanced on long thin legs, which move so quickly as they run that you can scarcely see them; while, when they fly, their long wings carry them lightly through the air, with their legs stretched out behind.

Fig. 40.

The Flamingo.
A duck-billed and web-footed bird among the waders.

What connection can there be between these active light little beings, and the broad-bodied web-footed swimmers? Go to the Zoological Gardens, and look at the Flamingo, with his long legs and curious curved beak. He is of the true swimming type, with his webbed feet and his sieve-like bill, with its rows of horny strainers like the geese; yet he feeds by wading in salt-water lakes and pools on the sea-shore, raking the bottom for food, and showing how a swimmer and a wader may once have had the same starting-point, before the one went out to sea, and the other in to shore. And then when we come back to our own little waders, and learn that they visit the sea, and feed upon the wet sands from the autumn to the spring, and then fly inland to build their nests in the damp meadows, feeding on earthworms, slugs, and insects of the land, we can see what an advantage this double life must be to them.

Notice, too, how shy and timid they have become from living among other animals, and watching for every danger. Try to get near one, and see how it will run on, turning its head hither and thither to watch, and at last will rise and be out of sight in no time. Or go and look for plover's eggs on the swampy grounds in our northern counties in the early summer, when

> "... from the shore
> The plovers scatter o'er the heath,
> And sing their wild notes to the listening waste."

The mother will no sooner see you than she will crouch down, running along a rut, and then move slowly away with a drooping wing as if wounded, hoping to make you follow her and pass by the little earthy hollow where her precious eggs are lying. The experience of life has made these little ground-nesting birds very intelligent, since they have had a land as well as a watery home, and the little moor-hen, which, like the rails and crakes, has taken entirely to a freshwater life in ponds, brooks, canals, and rivers, has learned to hide her nest so skilfully, and to dive and swim so cleverly, that even a trained water-spaniel often loses her when close upon her home.

And as the swimmers have their large birds in the albatross, so the waders too have theirs in the Herons, the Storks, and the Cranes. Who does not know how the storks fly in flocks to the sunny south in winter, and come back in the spring to build their nests in the chimneys of the houses of Holland and Germany, feeding on the banks of rivers, and in the fens on lizards, fish, frogs, and water-snakes; or how the cranes pass their summer in the stormy north, and their winter among the old ruins of Egyptian greatness? But the herons remain with us all the year, feeding on shrimps and crabs on the weed-covered shores, or more often in ponds and lakes upon frogs, water-rats, and fish. How patiently you may see a heron stand with his head slightly bent, still and motionless, till a fish passes by! Then quick as a flash of lightning, his head darts forward, impaling or seizing the prey in the strong beak, and he is off to eat it at his leisure. Thus he lives a solitary life all the year until the springtime, when he flies off to some group of lofty trees where for generations his family have built their nests, and, meeting with his fellows, piles up huge masses of sticks and grass among the tangled boughs.

And there the young herons are hatched and fed in the ancient heronry till they can perch and fly. For now among the waders we have come to birds that can perch, as we did among the swimmers (see p. 147). The heron has no longer the three-toed flat foot of the wader, with perhaps a slight spur behind, but a large fourth toe, with which he can grasp the bough; and as he flies across the country, uttering his strange harsh cry, often rising even higher than the hawks and falcons, and alighting on the top of some tall tree, few people would think of classing him among the waders, so like is he to those true land-birds whose life is spent in the air and whose home is in the trees.

THE FIRST KNOWN LAND BIRD

CHAPTER VII.

THE FEATHERED CONQUERORS
OF THE AIR.

PART II.—FROM RUNNING TO FLYING: FROM MOUND
LAYING TO NEST BUILDING: FROM CRY TO SONG.

SO the deserts and plains have their ostriches and cassowaries, the open ocean its albatrosses and its penguins,
the shores their ducks, gulls, and waders, and the little inland

pools and marshes their water-birds, which come there to build their nests and seek for food. Yet these are after all not by any means the larger portion of the bird world. It is in the woods and forests, the moors and pastures, on the solitary mountain peaks above, and in the snug valleys nestling below, that we find the myriads of song birds and game birds and birds of prey; of climbing birds such as the Woodpeckers; swiftly sailing birds such as the Swifts, cooing Wood-pigeons and cawing Rooks; terrible Eagles and Hawks, or sweet-singing Nightingales and Thrushes.

All these birds have had a very different education from that of the far-sailing albatross or the running ostrich. They have grown up in the midst of innumerable dangers; for enemies of all kinds—beasts and reptiles and other birds—live all round about them, making food scarce and destroying their young, so that of the millions born into the world thousands upon thousands perish every year before they grow up. We should expect, then, that these land birds would learn many devices for protecting themselves and their little ones. The guillemot can afford to lay her egg on the bare rock, for few animals can climb the high cliffs where she makes her home; and the penguin on her solitary island may lay hers in the mud on the ground. But the little lark must look carefully for high grass in which to build her nest, and the rook must weave a strong basket-work of twigs to make a home for her nestlings in the top of the high elm.

Moreover, the land birds cannot sleep safely on the ground, where weasels and stoats, foxes and wild cats, prowl by night in search of prey; they must take their rest on the boughs of the tall trees and cling on by their

toes even when they are in the deepest slumber. This
they could not do if they had the stumpy cushioned feet
of the ostrich, the webbed feet of the duck, or the flat
three-toed feet of the waders. It is the fourth toe turned
backwards, and growing very long in many of the perching birds, which gives them their grasp; while a special
muscle, beginning behind the thigh (*th*, Fig. 33, p. 125),
coming round over the front of the knee (*k*), and then
passing behind the heel (*h*), and on to the toes, keeps
them bent. Picture for a moment this muscle sending
its cords or tendons from behind the leg over the knee,
and then drawn back by the heel, and you will see that
the more heavily the bird sleeps, pressing upon its legs,
the more the knees will be bent forward, the tighter the
cord must be stretched, and the stronger the grasp will
be upon the bough.

Again, as to food, the land birds will be more closely
pressed than those which can at all times fish in the sea.
There is great scarcity of land food in the winter, while
in summer whole flocks of newly-born fledgelings are
clamouring for their daily bread. So we shall find that
every kind of eatable thing is turned to account, and we
have among land birds seed-eaters, vegetable-feeders,
and fruit-eaters; insect-devourers, and feeders on slugs
and worms and snails; and flesh-eaters which feed on
other birds, or on mice, bats, and larger animals; while
large flocks of birds of all kinds visit different parts of
the earth in the various seasons, going north in summer to build their nests, and south in winter, in search
of food. All these birds live chiefly in the air; while on
the ground there are the scratchers—fowls, partridges,
turkeys, and grouse, which rake out the hidden grains,

and rarely rise into the air except when they are frightened, or to roost on the trees at night. And between these ground birds and the true tree birds we have the doves and pigeons, some of which feed on fallen seed and grains, and others on fruit. And each and all of these birds have some difference in beak and claw, in their manner of nest building and rearing their young, and in their habits and ways, which enables them to make the most of their lives.

* * * * *

Even nest building does not come to all land birds by nature, and, as we shall see, it depends very largely on the habits and the structure of the builders. Thus the Partridges, and their relations the Pheasants and Grouse, lay their eggs in the thick grass of the meadows or among the heather, and at most sometimes scratch together a few dry grass blades for a bed. In this they remind us much of the ostrich family, which also scrape a hole in the ground for their eggs and scratch food for their children; and in fact there is a group of curious heavily-flying birds, called Tinamous, in South America, which are so like quails and partridges on the one hand, and ostriches on the other, that they lead us to wonder whether it was not from the ancestors of such birds as these in ancient times that the heavy running birds started on one road, while the lighter and swifter birds took to the wing.

The wings of all the scratching birds are even now short and round, and their flight is feeble. Their chief home is on the ground, where they crouch among the thick herbage when the keen-eyed hawk is hovering overhead, never taking to their wings till no other chance is left them. The mother partridge runs many dangers as she sits upon her dark-coloured eggs in some sheltered spot, for weasels and stoats will attack her and steal her eggs if she leaves them for a moment, or kill her herself if they can take her unawares in the dark night. She could never hope to rear her young ones if they did not come out of the egg well covered with down, and able to run and pick by her side while she and the father scratch the ground with their short blunt claws to get ant-cocoons, and later on worms and insects for them.

Yet so well does scratching answer, in getting at buried food such as other birds cannot find, that there are a large number of these ground birds all over the world. The Guinea fowls of Africa, the spurred Peacocks, Pheasants, and Jungle fowls of India (from which last our tame fowls probably come), the wild Turkeys of America, the Quails which live in all parts of the old world from Australia to England, and the Ptarmigans of our northern countries, which put on their white plumage in winter—all these show how advantage has been taken of every nook in which ground birds could find shelter. We find them hiding in thick jungles and shady woods, or even in open ground among high grass and corn, scratching mother earth for their daily food; washing not in water but actually in the dust, by rolling in it, and then shaking it off; escaping many dangers by wearing a plumage very much the same in colour as the different grasses and leaves among which they hide; and feeding on insects, worms, and seeds, and whatever they can find upon the ground or under it.

Fig. 41.

Brush turkeys[98] and their egg mounds.

And when we travel far off to Australia, we find ground
birds which do not even sit on their eggs, nor take care of
their young, but leave them as reptiles do to be hatched in
the sun. The Brush-turkeys and Megapodes of Australia
and the islands near, and the Maleos of Celebes—all of them
scratching birds—come out of the thick jungle and lay their
brick-red or pale-coloured eggs on the shore, never taking
any more notice of them. The maleos simply scratch a hole
in the sand and bury the eggs, the brush turkeys and megapodes[99] scratch together all kinds of rubbish and dead leaves,

98   Talegallus.
99   Megapodidæ or large-footed birds.

carrying them in their long curved claws, and adding them to the heap till they have made a mound sometimes more than seven or eight feet high, and twenty feet across at the base; an astounding size, when we consider that the brush turkeys are not nearly as large as a good-sized turkey, and the megapodes not larger than hens. It is to these mounds that the mothers come about every ten days, and lay an egg *upright*, till each has laid eight or nine, and then she comes no more; but after many weeks the little chicks work their way out fully fledged, and fly away to get their own living. The probable reason, Mr. Wallace tells us, for this curious habit of mound-building, is that the eggs are so large that the mother can only lay one every ten days, so that if she sat upon them she would be worn out with fatigue and want of proper food before they were all laid and hatched.

* * * * *

We see then that the scratching birds live nearly all over the world, yet, no doubt, it is a disadvantage to them that in their ground life they have become so heavy that they cannot fly so lightly or so far as their near allies, the pigeons, which, like them, feed on the ground. For the Pigeons have already made many steps forward in life. Their wings are strong, so that they can fly for great distances; their toes are slender and well fitted for perching; and though it is true that our tame pigeons and the wild rock-pigeons from which they are descended do not build nests, but lay their eggs in dovecots or church towers, or, if they are wild, in holes in the rocks, yet the beautiful blue-gray wood-pigeon, with her pure white collar and soft cooing note, builds a nest in the trees—

Fig. 42.

Wood-pigeon on her nest.

"The stock-dove builds her nest
Where the wild flowers' odours float;"

though it is but a rough one, made, as well as her weak feet
and bill can do it, of a few stout twigs, laid so loosely that
her two little white eggs may be seen from below, and even
sometimes fall through.

Yet, though but a beginner in nest building, she is a true
tree bird, and her little ones are born naked and helpless,
far out of reach of the ground, and must be fed and cared
for till they can fly. So she feeds them with infant pap from
her own mouth. The "crop" or bag in which the partridge
or hen stores the grain she picks up is large and single; but
the pigeon has two bags, one on each side of the throat, and

when she is feeding her young these bags secrete a large quantity of milky fluid, which, mixing with the tender shoots she has pecked off in the spring, or with the oily seeds she has gathered for her autumn brood, makes a soft food, which she pours into the mouths of her nestlings till they fly and feed themselves.

In the pigeons, then, we are gradually rising from the ground birds,—where the father generally has many wives[100] and the young ones run as soon as they are hatched,—to the tree birds, where father and mother, taught by the helplessness of their brood, share the cares of nest building and the pleasures of love. Even the pigeons did not all at once become tree birds, for we have them in all stages now from the ground to the air. Many years ago, in the island of Mauritius, there were heavy flat-breasted pigeons, the Dodos, which lived entirely on the ground without the power to rise, so that when the Dutch settled there, bringing rats with them in their ships, the Dodos soon fell victims to the intruders, and now there is not one left. Again, in New Guinea now, there are ground pigeons which fly heavily and slowly, and only go to the trees to roost. Then come our own tame pigeons, the rock-pigeons, and the stock-dove which builds in boles in the trees; and then our wood-dove and his relations, with their rude nests and their mixed food of grain and grass. And among these are the wonderful long-winged passenger pigeons[101] of America, which fly in flocks of hundreds of thousands through Ohio, Kentucky, and Indiana, in search of nuts and seeds, breaking down the boughs of the trees by their weight where they alight, and then darkening the whole sky as they start off again in a succession of vast

100   Partridges, quails, and some others are exceptions, and pair.
101   Columba migratoria.

multitudes to another forest where beech nuts, acorns, and chestnuts are plentiful, or to the rice-grounds of Carolina, to take their fill.

And, lastly, we come to the beautiful green fruit-eating pigeons of India and the East—the feeders on nutmegs and palm-fruits and juicy berries of all kinds. These are true tree birds, difficult even to find, so like are they to the colour of the leaves; yet they still build the loose untidy nests of their kind.

Nor need we wonder at this, for fine nest building requires both strength and delicacy in the bill and feet; and the next group of birds escapes it altogether by finding or making holes in trees and banks, and lining them with moss or leaves. This group is the Climbers, which come, as it were, between the ground birds and birds of active flight, for they clamber swiftly up the trunks and over the branches of trees in search of fruits and insects, seldom going down to the ground, but flitting from tree to tree to find fresh hunting grounds.

What is that green object, about as large as a small squirrel, which we see mounting the trunk of one of the elm trees, as we lie resting on the moss in some quiet wood? There it goes, dodging now to this side, now to that, with its head well lifted and its stiff tail bent against the trunk. It is the green woodpecker at his work. His long large feet, with toes divided *in pairs*, two in front and two behind, take firm hold of the tree with their sharp claws; his breast, which is flatter than that of most birds, lies close against the bark, as he mounts by a number of rapid jumps, which are made by pressing his strangely stiff horny tail against the trunk, while he hops forward with both feet, making a slight rustling noise, and moving so fast that it is difficult to see how he does it.

Fig. 43.

The great green Woodpecker.[102]

Now he pauses; it is to try a suspicious place in the bark, and tapping it with his beak he finds that it gives a hollow sound. This tells him at once that it is rotten, and there is an insect within; and pecking a hole with rapid blows of his chisel-like bill, he inserts his narrow bill, and darts a long gluey tongue, with barbed tip, into the dark passage, bringing out the intruder, which is swallowed in a moment. A

102   Gecinus viridis.

strange tongue this is of the woodpecker, for it has two long bony branches at its roots, and each one is like a bow bent under and round the back of the bird's head, and as these bows are tightened or slackened by the slender muscles the tongue is drawn in, or thrust out to an extraordinary length. Moreover, it has at its tip a horny covering beset with tiny barbs, and every time it goes back to the mouth these are bathed in gluey slime to catch the next insect it may meet. Nor is the woodpecker obliged always to drill for his food. The tiny ants, as they wander up and down the trees, the beetles and bees settling on the branches—all may fear this gluey weapon, for all alike disappear within the long thin beak.

And now, perhaps, our friend has flown to another tree, and is some way up it. Where is he gone? Climb up and look, and you will find a small round hole, small outside but not inside, for the woodpecker has hollowed out the soft rotten wood, and within, if it be early summer, the mother is still sitting upon five or seven pure white eggs, out of which the naked little ones will soon creep. He is a clever fellow the woodpecker, but he is by no means the chief or most conspicuous of the climbers, for in this group we have some of the most gaudy and remarkable of birds. The brilliantly-coloured Barbets, the gaudy-headed Toucans, with their clumsy bills and long barbed tongues, and the gorgeously-tinted Parrots and Parroquets, with their soft fleshy tongues so well adapted for speech, are all climbers, with toes divided two and two, and they wander about the trees of South America and the East, feeding on fruits and seeds.

Where in any other part of the animal kingdom can we find so many brilliant colours crowded together as in the plumage of birds, and especially in birds of tropical countries? The large land animals cannot afford to wear such bright coats lest they should attract their enemies, nor can even birds often put on gay plumage in our northern climates, where the trees are bare for half the year. But in warm sunny latitudes, where the trees are always green and the foliage thick and heavy, and where brilliant fruits and flowers often peep out among the leaves, the gaily-coloured birds can wander in comparative safety, and even the gaudy parrots are not easily detected as they clamber from bough to bough, using not their *tail* like the woodpecker, but their *beak*, as a third foot to hold on by as they climb.

None of these birds build nests; indeed, they could hardly do so with their clumsy beaks and thick heavy feet; they either, like the ground parrots, put together a few leaves in hollows of the earth or in ants' nests; or, like the fruit-eating parrots and toucans, they lay their eggs in tree-holes, where the bright-coloured mother is safely hidden till she is set at liberty again. Even the cuckoos which, though they are climbers, have taken much more to the wing than their associates, sometimes avoid the trouble of nest building by laying their eggs in the nests of other birds, as our own spring visitor always does. Some of them, however, in America and elsewhere, have contracted better habits, and build very respectable nests of their own.

Indeed, we shall now soon begin to make progress in nest building, for the next group of birds, which *dart* at their food with wide-gaping mouth and seize it on the wing, have among them many clever little architects. It is true our English kingfisher builds in holes on the river bank, lining her nest with fishes' bones, and the Nightjar (wrongly called the goatsucker), with her wide-gaping mouth, lays her egg on the ground. But both these are lowly Darters, for the

Fig. 44.

The Kingfisher.[103]

kingfisher sits on a bough close above the water, and pounces down upon the fish or water-insects; and the lonely nightjar, with her strange wailing cry, flits among the bushes in the twilight, or often even creeps after her prey.

Neither of these birds can compare in flight with the Swifts, as they dart upon the wing from some high pinnacle to collect a mouthful of insects, and come back to eat them, nor to the lovely little Humming Birds of America, which poise themselves so deftly on the wing, while their

103   Alcedo ispida.

slender bill searches the long-tubed flowers for insects or
seizes these as they pass. These living jewels of nature build
beautiful and delicate nests of leaves and grass and spiders'
webs interwoven like fairy cradles; while the swift makes
a far stronger home of hair and feathers, grass and moss,
glueing them together with saliva[104] from his mouth, and
fastening them under the eaves or on the top of some tall
waterspout. It is easy to see why the swift chooses such lofty
spots, for his slender weak toes are ill-fitted for standing on
the ground, and he rises with great difficulty when once he
has alighted there, but from a height he can drop easily on
to the wing, and skim the air for his food.

Now the swift, which visits us only in summer to build
his nest, when insects are plentiful, and spends the rest of
his time in Africa, is a type of a whole army of birds, lovely,
bright, and gay, with short weak feet, long wings, and a
gaping mouth surrounded by bristly hairs, which swarm
in hot countries where insects are to be found all the year
round. Among these are the beautiful little Bee-eaters and
Rollers of the East and Africa, which revel in insect food, and
sometimes visit us in the summer, coming over to the south
of Spain, or even, in the case of the rollers, as far north as
Sweden; while in South America the dull-coloured Puff-birds,
the brilliant Jacamars with their metallic-looking feathers,
the delicate little Todies, the bright green Motmots, and the
lovely Humming-birds, swarm in countless numbers, hiding
among the dense foliage, or darting in the bright sunshine
after bee or butterfly, or other unwary insects.

* * * * *

But we must not pause too long among these smaller

[104]   The Indian and Chinese edible-nest Swiftlets (Collocalia), make their nests
entirely of this saliva, and they are eaten by the natives.

groups of birds, for the multitude of perching birds, which form nine-tenths of the whole bird kingdom, await us with their delicate nests and their happy family life. Ah! now we are really coming to nature's feathered favourites, for what can be sweeter than the song of the nightingale, the skylark, or the thrush? or more touching than the fact that the young ones learn from their father the loving notes; that they, in their turn, may be able to woo and win some gentle mate to share their nests and bring up their young ones? It is for this that they have gained that wonderful singing instrument which they have deep down in their throat. For they do not produce their sounds as we do, just below the back of the mouth, but at the lower end of the windpipe, just where it divides into two branches, one going to each lung. There, where the rush of the air is strongest, is found a complicated apparatus, moved by a whole set of muscles, upon which the little fellow plays, and seems never to be exhausted, so much air has he in all parts of his body. And as the song pours through the windpipe there again he can help to give it its soft mellow tones, for while in hoarse-crying birds, like the sea birds and the waders, this tube is long and stiff, in the sweet singing birds it is short, and the bony rings composing it are thin and far apart, with soft delicate membrane between them, which can be shortened or lengthened to modulate the tones. And so we hear them in the springtime pouring forth their full tide of song to tempt a young wife to come and help them to build a nest; or, in the full pleasure of success, trilling out their delight in the warm bright sunshine, and calling on all the world to be as happy as they.

Yet it is not by any means all the perching birds which have this wonderful gift of song. Even among our own birds, the jay, the crow, the raven, and others, use their musical instrument for talking in a way that is no doubt useful to them, but scarcely pleasant to hear; and in America there is a whole group of songless perching birds—the bright coloured chatterers, the fly-catching tyrant-birds, the American ant-thrushes, which have not even developed a true singing instrument in their throat, and only utter shrill or bell-like cries. Yet they all build nests and cherish their helpless young ones; and so large and varied is the group of perching birds, whether in the Old or New World, that they fill all the stray nooks and corners of bird-life, often imitating the habits of the other smaller groups so as to get at food of all kinds. Thus, while the Finches with their delicate matted nests, the Warblers, and a large number of the smaller birds, lead a true tree and bush life, feeding on fruits and insects, the Thrushes, Blackbirds, Crows, Redbreasts, and Larks are *ground-feeders*, which, though they do not scratch with their feet like the partridges, turn up the ground with their bills and pick out the worms and grubs.

For this reason the Song-thrushes love to build their nest of twigs and moss lined with soft wood chips, in some thick hedge near to a meadow or garden, where they can drop down and pull up the unfortunate worms before they have gone home underground after their nightly rambles, or pounce upon unwary snails, and, taking them in their beak, crack the shell upon a stone, and carry off the dainty morsel to their brood; while the Lark, with her long hind toe, so well fitted for walking, hides her nest in a furrow on the ground; and the greedy cunning Magpie, feeding, as she often does, on young animals, seems to fear the same fate for her own brood, and builds a large egg-shaped dome of thorny branches, with the thorns sticking out on all sides, and lined with mud and soft roots, leaving only a small hole for a door. Lastly, the sagacious Rooks, though ground-feeders, build strong homes which last from year to year, in the top of the high elms, and set out in companies in the early morning to their feeding grounds.

Fig. 45.

Nest of the Common Wren.[105]

105   Troglodytes parvulus.

Then, as there are ground-feeders among the perchers, so, too, there are *climbers*, for the Creepers, the Wryneck, and the Nuthatch, run up and down the trees, feeding on insects and nuts, which the nuthatch breaks so cleverly with his beak; and we might almost fancy them to be first cousins to the woodpeckers, if it were not for their three toes in front and long claw behind, and their short thick beak and tail. Even the little Wren, with her cocked-up tail, imitates the climbers as she creeps through the hedges and underwood, though she is a true perching bird, and builds one of the most perfect of nests of moss and grass, woven into the shape of a ball, with a tiny hole for a door. Then, to match the *darting* birds, we have the Swallow and the Fly Catcher which follow insects on the wing, so that the swallow and swift were long confounded together, though the skeleton of the swallow shows that it belongs to perching birds. Again, the Shrike imitates the birds of prey, feeding on small mice, reptiles, and birds, and impaling them upon a sharp thorn while he tears them to pieces with his beak. Yet he is a true percher, singing as beautifully as many of the smaller birds, and he is even said to use his power of song to lure victims within reach. Lastly, and perhaps most curious of all, the little Dipper or Water-Ouzel, with his clear loud song, and his structure so like to the thrushes, has actually taken to the habits of water-birds, and dives into the depths of the river, running along upon the bottom and feeding on water-snails and water-insects.

All these we find among English birds; and if we had space to speak of other countries, we should find the same history there, for the more we study bird-life the more we find that these Perchers are its highest types, and have learned to make the most of their kingdom. It is they who build the most per-

Fig. 46.

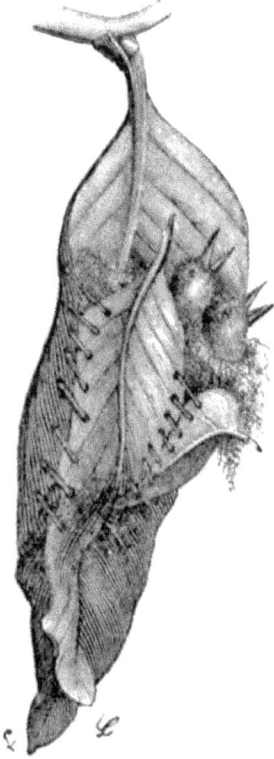

NEST OF THE TAILOR-BIRD[1]
OF INDIA OR CHINA.

1   Orthotomus sutorius.

fect nests, from the rough strong basket-work of the crow or the magpie, to the wren's thickly-woven ball, or the finches' matted cups; while in America the Hang-nests weave their lovely pear-shaped homes, and suspend them like fruit from the tips of the branches; and in India and China the Tailor-birds actually sew leaves together with cotton fibre or cobweb threads, which they draw through with their slender bill and strengthen with saliva.

The smaller the bird and the more delicate its feet and bill, the more closely woven, as a rule, is its nest. Yet all are built with care; the mother bird, as a rule, choosing the position and laying the twigs, while the father helps her to collect the materials. So rapidly do these little creatures work, that among our smaller English birds the early morning sees the work begun, and by evening it is ended. Other birds are longer, according to the amount of material they have to collect; but all labour industriously till the cradle is finished, and then begins the laying, the sitting, the tender care of the mother for her little ones, and of the father for his wife and brood.

* * * * *

And indeed there is much need both of skill in nest building and of watchfulness for many a long day after, for if
the perchers are the highest, they are not by any means

Fig. 47.

The Eagle bringing food to its young.—(*From a
coloured lithograph by Keulemann.*)

the strongest of birds; and while they feed on insects and smaller creatures, they have to guard their little ones with anxious care against the larger birds of prey which rule as masters in the higher regions of the air. It is on rocky pinnacles and in the clefts of inaccessible heights among the mountains that we must look for the nests of the Eagle, the Vulture, and the Falcon. Strong, powerful, and untiring in flight, they sail majestically high up in the air, not to sing a joyful song like the lark, but with piercing eye to search every corner for miles around, for animals of all sizes, from the dead ox or mule to the tiny living mouse or bird, which can serve for a meal.

It needs only a glance at them to see that they are strong destroyers, with their powerful wings, their sharp hooked beaks, their long toes with pointed claws, and their strong muscular thighs; and because most men admire strength and power, we call such birds *noble*, though their nobility chiefly consists in loving their little ones, and asking neither pity nor shelter from others, as they themselves are pitiless in return. Those which we are apt to like the least, the carrion-feeding Vultures of hot countries, are really the most useful and harmless, for they feed chiefly on dead animals and clear the land of carrion; and for this reason neither their beak nor their claws are as strong as those of the fighting birds. But though they are grand in flight they are but repulsive-looking birds when compared with the lordly eagles. The beautiful Golden Eagle of Europe, with its dark plumage and the golden sheen on its back and tail, is indeed a splendid object, as

> "He clasps the crag with hooked hands,
> Close to the sun in lonely lands,"

or still more, as he sweeps along with steady flight, circling round and glancing with searching eye over the plain beneath. Suddenly his attitude changes; he closes his wings, and, head downwards, drops to earth slantingwise with a rushing noise, seizing in his claws the startled fawn as it dashes by at full speed, the frolicking rabbit darting into its hole, or the terrified bird upon whom his choice has fallen. Then, with a powerful stroke he rises up again, and is lost to sight as he soars aloft and regains the rocky peak where his eyrie is built and his children are clamouring for food.

So, too, the dexterous Falcon swoops upon his prey swift as an arrow, his pointed wings striking the air, and then closing at once upon his body, while his long rounded tail guides him in his flight. Who would think that such a powerful and bold robber could have anything in common with the soft feathered owl which sits blinking its large eyes in the hollow of the tree till the twilight falls? And yet the Owl, with very little change in structure, has become as fitted to follow prey at night as the falcon is by day—

"What time the preying owl, with sleepy wing,
Sweeps o'er the cornfield, studious."

The soft, round, broad wings, which would serve badly for striking a quarry from on high, are exactly fitted for gliding in the silence of the night, as, guided by wide open eye and ear, he skims over the fields or round the stacks in the yard to pounce noiselessly upon the unwary mouse or to seize the flying beetles and bats. Then the sharp claws appear quickly from under the downy feathered feet, and clutch the smallest prey with needle-like precision; and away the owl flies to his nest, so quietly that even the other animals close by are not alarmed, but in ignorant security remain till he comes to strike again.

And as the day and the night by land have their relentless freebooters, so the sea too has its eagle king; for the Osprey, with its nest on a high rock, hovers over the open sea, and, dashing into the deep, returns with a large fish in its claws; and, as it tears the flesh from under the glittering scales, reminds us that there is no spot on the earth in which some bird does not seek its prey.

We have now in very brief outline followed the feathery tribe from the flightless penguin to the boldly-soaring eagle, the king of the air. Those feathers which in the swimming bird are scarcely more than finely-divided scales, and in the ostrich mere loose nodding plumes, have become in the albatross, the vulture, and the soaring falcon, flying instruments of such power and strength that the earth and the water are as nothing to them compared with the free ocean of air; while even the tiny graceful swallow flies for hundreds of miles to its winter home.

Indeed, we have here one of the great secrets of bird success; for while most animals must roam within limited districts, and get their food there as best they can, thousands and tens of thousands of birds set off, when the colder weather makes food scarce in any one region, and travel hundreds of miles to more genial climates, where insects are still to be found, and the trees are still covered with fruit and leaves. How strange it is to think that while we are making the best of our winter, the swallow has taken her unerring flight to Africa, the swans and cranes have long since made their southward journey, and myriads of small birds have gone in search of food and warmth, to return next spring as certainly to their old haunts, where they can breed in cool and comfortable quarters!

If we could only get the birds to tell us how they have learned the routes they take, and by what rules they are guided! One thing we know, that each kind of bird makes its nest in the coldest region which it visits, and where, at the time its young brood are ready, insect and other life is abundant; so that while the wild duck and goose, the woodcock, snipe, and field-fare, go to the far north to lay their eggs, and come to us in the sharper weather to feed when there is nothing but ice and snow in the home they have left, the swallow, the cuckoo, the swift, and the wheatear, on the other hand, visit us in the spring to build, and when autumn comes on take their flight to Africa and the East; and even many of the song-thrushes and robin-redbreasts which remain with us in England start off from Germany to warmer climates. Others, again, such as some of the Reed-warblers, the Stint, and the Ortolan Bunting, only make our island a house of call between the arctic regions where they breed in the summer when mosquitoes are swarming there, and the south where they winter after flying thousands of miles.

It would take too long to discuss here why and how they go, even if we knew it with certainty; but it is most probable that their ancestors first learned the routes now taken when Europe and Africa had not so wide a sea between them, and we can see that it must be a great advantage to be able to travel from climate to climate, so as to find a plentiful table spread at all times of the year; while they may return to the north to breed, not merely because there is food there, but also because in still earlier times, when the northern countries were much warmer than they are now, they doubtless lived there altogether, and, though now obliged to go south, have never lost the tradition of their old home.

Thus the birds, with their feathery covering and powerful

wings, have left their early friends, the reptiles, far far behind. Taught by their many dangers, many experiences, and many joys, they have become warmhearted, quickwitted, timid or bold, ferocious or cunning, deliberate as the rook, or passionate as the falcon, according to the life they have to lead; or, in the sweet tender emotions of the little song-birds, have learned to fill the world with love and brightness and song. If mere enjoyment were all that could be desired in life, where could we expect to find it better than in the light-hearted skylark as she rises in the early summer morning to trill forth her song of joy, or in the happy chuckle of the hen as her little ones gather around her.

Yet we cannot but feel that, happy as a bird's life may be, it still leaves something to be desired; and that, with their small brain and their front limbs entirely employed in flying, they cannot make the highest use of the world. The air they have conquered; and among the woods and forests, over the wide sea, and above the lofty mountains, they lead a busy and happy existence, bringing flying creatures to their highest development, and showing how Life has left no space unfilled with her children. Yet, after all, it is upon the ground, where difficulties are many, conditions varied, and where there is so much to call for contrivance, adaptation, and intelligence, that we must look for the highest types of life; and while we leave the joyous birds with regret, we must go back to the lower forms among the four-footed animals, in order to travel along the line of those that have conquered the earth and prepared the way for man himself.

HOME OF THE EARLIEST
KNOWN MARSUPIALS

CHAPTER VIII.

THE MAMMALIA OR MILK-GIVERS.

OUR backboned animals have now travelled far along the
journey of life. The *fish*, in many and varied forms, have

taken possession of the seas, lakes, and rivers; the *amphibia*, once large and powerful, now in small and scattered groups, fill the swamps and the debateable ground between earth and water; the *reptiles*, no longer masters of the world, but creepers and skirmishers still holding their own in many places either by agility, strength, or the use of dangerous weapons, swarm in the tropics, and even in colder countries glide rapidly along in the warm sunshine, or hide in nooks and crannies, and sleep the winter away. And the *birds*,—the merry, active, warmhearted birds,—live everywhere, making the forests echo with their song, rising into the heights of the clear atmosphere, till the world lies as a dim panorama below them, crowd the water's edge with busy fluttering life, and even wander for days and weeks over the pathless ocean, where nothing is to be seen but sky and water.

Yet still the great backboned division is not exhausted; on the contrary, the most powerful if not the most numerous group is still to come; that group which contains the kangaroos and opossums, the dreamy sloths, the night-loving moles and hedgehogs, the gentle lemurs and the chattering monkeys, the whales, seals, and walruses for the water; the herds of wild cattle and antelopes, of noble elephants and fleet horses, for the forests, mountains, and plains; and the ferocious beasts of prey, which make these gentler animals their food; while last, but not least, comes man himself, the master and conqueror of all.

Where, then, shall we look for the beginning of this vast multitude of warm-blooded, hairy, and four-limbed animals? If we turn back to the past, we get but little help; for though in that early time, when huge reptiles overran the world and swam in the waters, we find small animals (see Fig. 48), probably of the marsupial or pouched family, living in the forests, yet even if these were the earliest of their race, which is not at all likely, they would tell us very little about the beginning of the milk-givers, since only their lower jaws remain, and we can only guess at their relationship by these having that peculiar inward bend which we still find in all pouched animals.

Fig. 48.

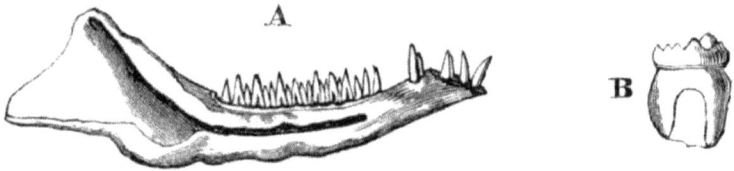

A, Jaw of Dromatherium; B, Tooth of Microlestes; both milk-givers, probably marsupials, found in beds of the same age as those containing the ancient swimming lizards.

No! for the few scattered facts about the lowest *mammalia* or milk-giving animals we must inquire of our own day, to learn something as to the causes of their success in life. And first let us notice two important changes which give them an advantage over other backboned creatures. We have seen that, as we have gradually risen in the scale of Life, parents have taken more and more care of their eggs and their young ones. Among the boneless animals which we studied in *Life and her Children*, it was not (with very few exceptions) till we reached the clever, industrious, intelligent insects, that we found them taking any thought for the weak and helpless infants. There we did find it, for insects in their own peculiar line stand very high among animals; when, however, we turned back again to begin with the first feeble representatives of the backboned family, we found the fish casting their eggs to the bottom of the sea, or on the pebbly gravel of a flowing stream, and, as a rule, taking no more thought of them. The tiny stickleback with his nest, and the lumpsucker watching over his young ones, were quite exceptions among the finny tribe. So it was again with the frogs, so with the reptiles (the turtles, lizards, and snakes), whose eggs, even when carefully buried by the mother, are often devoured by thousands before the little ones have a chance of creeping out of the shell. But when we come to the birds, there, as with the insects, we find parental care beginning—the nest, the home, the feeding, the education in flying, in singing, in seeking food, the warm-hearted love which will risk death sooner than forsake the little ones.

Yet still these same little ones have many perils to run even before they break through the shell. In spite of their parents' care, more eggs probably are eaten by snakes or weasels, field-rats, and other creatures, than remain to be hatched; while, even if they escape being devoured, the eggs must not be allowed to grow cold; and should the parents be too long away or be scared off the nest by some enemy, or should a damp cold season spoil the warm dry home, the young bird is killed in the egg before it has ever seen the light.

It is not difficult to see, therefore, that if the mother could carry the egg about with her till the little bird was born, as we found our little common lizard doing (see p. 105), it would be much safer than when left in the nest exposed to so many dangers.

Now something of this kind takes place with all that great group of animals we are going to study. The cat and the cow, as we all know, do not lay eggs as birds do; but the mother carries the young within her body while they are going through all the changes which the chicken goes through in the egg. Thus they go wherever she goes, the food which she takes feeds them, and they lie hidden, safe from danger, till they are born, perfectly formed, into the world. Nor is this all; for when at last her little ones see the light, the mother has nourishment ready for them; part of the food which she herself eats is turned into milk, and secreted by special glands, so that the newly-born calf or kitten is suckled at its mother's breast till it has strength to feed itself.

These two advantages, then,—namely, that the young have no dangerous egg-stage, but are sheltered by their mother till they are perfect, and that their mother has milk to give them for food,—at once divide the *Mammalia* or milk-giving group of animals from the rest of the backboned family.

But how will this help us to learn where that great group begins? Is it possible that such creatures as these can have anything in common with reptiles and birds? To answer these questions we must travel to a part of the world which has long been separated from the great continents of Europe, Asia, Africa, and America, and where the low and feeble milk-giving animals had a chance of still keeping a place in the world.

Take a map and look at Australia, New Guinea, and Tasmania, and you will see that they are separated by a number of scattered islands from the great continents, which are not only large in themselves, but are all nearly joined together, with only narrow straits dividing them. Moreover, Australasia stands even more alone than appears at first sight; for Mr. Wallace has pointed out that a very deep sea separates New Guinea and Australia on the one hand from Borneo and China on the other; so that the land might rise several thousand feet, and yet the Australasian islands would not be joined to the great continents.

Now, if the milk-givers once had feeble beginnings, and gradually branched out, as the ages went on, into all the many forms now living, it is clear that on the great battlefield of Europe, Asia, Africa, and America, the first poor weak forms would gradually be destroyed by the stronger ones that overran these great continents. They would be crushed out, as so many of the reptiles and newts and fishes had been before them; and only their bones, if any remained, would tell us that they had once lived. But if some of them could find a refuge in a domain of their own, where after a time they had a good open sea between them and their stronger neighbours, they might have a chance of living on and keeping up the old traditions.

And this is just what we have reason to believe has been their history; for it is exactly in Australasia that we find that curious group of pouched animals, the kangaroos and other *Marsupials*,[106] as they are called, which are different from all the other milk-giving animals in the world, except the opossums of America, whom we shall speak of by-and-by.

And together with these marsupials we also find the simplest milk-giving animals now living. Come with me in imagination to a quiet creek in one of the rivers of East Australia. It is a bright summer day, and the lovely acacias are hanging out their golden blossoms in striking contrast to the tall graceful gum-trees and dark swamp oaks in the plain beyond. Come quietly, and do not brush the reeds growing thickly on the bank; for the least noise will startle the creature we are in search of, and he will

106   *Marsupium*, a pouch.

dive far out of sight. There he is, gently paddling along among the water plants. His dark furry body, about a foot and a half long, with a short broad tail at the end, makes him look at first like a small beaver. But why, then, has he a flat duck's bill on the tip of his nose, with a soft fold or flap of flesh round it, with which he seems to feel as he goes? Again, he has four paws, with which he is paddling along; but though these paws have true claws to them, they have also a thick web under the toes, stretching, in the front feet (C Fig. 50), far beyond the claws, yet loose from them, so that while it serves for swimming it can be pushed back when the animal is digging in the ground. His hind feet have a much shorter web, and a sharp spur behind, like that of a game cock.

And now, as this animal turns his head from side to side you can see his sharp little eyes, but not his ears, for they are small holes which he can close quite tightly as he works along in the water, pushing his bill into the mud of the bank, just as a duck does, and drawing it back with the same peculiar jerky snap; for he too has ridges in his beak like the duck family, through which he sifts his food; while, at the same time, he has in his mouth eight horny mouth-plates, peculiar to himself.

What, then, is this four-footed animal with a beaver's fur and tail, and teeth in his mouth, and yet with a duck's bill and webbed feet? He is the lowest and simplest milk-giving animal we know of in the world—the duck-billed Platypus or Ornithorhynchus, called by the settlers the Water-mole.

Fig. 49.

The Duck-billed Platypus[107] swimming and rolled up, with its underground
nest laid open behind; on the right hand bank is an Echidna.[108]

If we could search along the bank we should find, somewhere below the water's edge, a hole, and again, a few feet
back on the land, another among the grass and reeds; and
both of these lead into a long passage, which ends in a snug
underground nest—a dark hole lined with dry grass and
weeds—where in the summer time (about December) we
should find the mother platypus, with two or four tiny naked
young ones, not two inches long, cuddled under her. How
these little ones begin life we do not know. The natives talk
about finding soft eggs like those of reptiles; but it seems
more likely that these eggs break just as they are laid, like

107   Ornithorhynchus paradoxus.
108   Echidna hystrix.

those of our common lizard (see p. 105), and the naked little ones come out alive into the nest.

And how are they fed? Their mother has no teat, like the cow, to put into their mouth, for she is a very primitive creature; only in one spot amid her fur are a number of little holes, and from these she can force out milk for them to drink as they press against her with their soft flat bills. So here, in a dark underground nest, away from the world, because she cannot, like the higher animals, carry her little ones till they are perfect, the duck-billed platypus, which may well be called "paradoxical" (see Fig. 49), enables us to picture to ourselves how, in ages long gone by, mothers first began to feed their little ones with their own milk.

And now, perhaps, you will be struck by this animal's likeness to a bird, especially when you hear that the little baby water-moles have a soft horny knob on their nose, just where young birds have a hard knob for breaking through the shell; and you will ask if milk-giving animals came from birds. Not at all; young tortoises, too, have such a knob, and so have crocodiles; and, moreover, these duck-billed moles have many parts of their skeleton, especially the shoulder bone and the separate

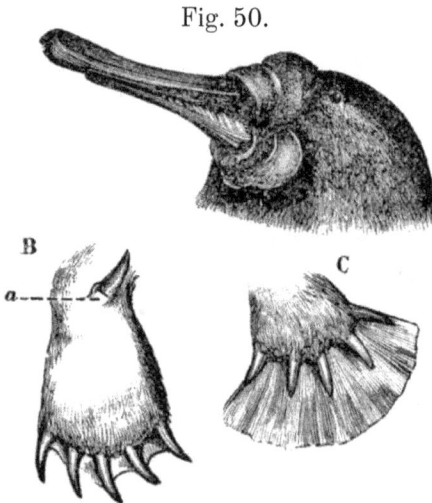

Fig. 50.

A, Head of Ornithorhynchus, showing serrated bill; B, Hind foot with claw a, found on the males only; C, Webbed fore foot.

bones of the skull, very like our living reptiles, and still more like some which lived in ages long gone by.[109] And yet at the same time they differ essentially both from reptiles and birds in many points besides those we have been able to mention, and in one in particular, which we can understand now we have studied these groups, namely, that the platypus, like all milk-giving animals, *is without that curious quadrate bone* (*q*, Figs. 23 and 33) *which we find in all reptiles and birds.*

Now, notice the frog, which is an amphibian and therefore lower than the reptiles, has not got this quadrate bone, though his companions the newts have; and he seems to tell us that among those old amphibians which roamed in the coal-forests of ages past, there must have been some which,—while they had that great mass of cartilage which imperfect, unborn, milk-giving animals have even now, out of part of which this bone is formed,—yet never went so far as to have the bone itself. If this is so, then here at last, in the distant past—so remote that we cannot even guess how long ago it may have been—we have a point from which the earliest ancestors of the milk-giving animals may have gone off in one direction, and those of reptiles and birds in another. And this would explain how it is that they have so many points in common, while yet the mammalia are without that special bone and other characters which are found both in reptiles and birds.[110]

Be this as it may, here is our lowest mammalian form,

109   Professor Owen has described a reptile from the Trias of Africa, and Professor Cope another from the Permian of Texas, both having characters closely resembling the Platypus.

110   This argument, which can only be stated very roughly here, must not be supposed to rest merely on the quadrate bone, though this is the easiest point to illustrate popularly. I am deeply indebted to Mr. W. Kitchen Parker for a whole flood of light thrown on these early forms, and only regret that I have neither skill nor space to do justice to his graphic illustration of a subject of which he is pre-eminently master.

and he has a relation, the Echidna, very like him in many respects, but who has made a decided step forward; for on the sandy shores and in the rocky gorges of Australia, creatures about a foot long, covered with prickly spines like hedgehogs, and called by the settlers "Porcupine Ant-eaters" (see Fig. 49), shuffle along in the twilight, thrusting out their long thin tongues from the small mouth at the end of their beak-like snout, and feeding on ants and ants' eggs. These do not belong, however, to the real ant-eater family, but are near relations to the platypus; and they are well protected by their spines in the battle of life, for when attacked they either roll themselves up into a ball like a hedgehog, or burrow down into the sand so fast that they seem to sink into it, leaving only the points of their prickles sticking out to pierce the feet of their enemy. Now these creatures have a little fold of skin under their body, which forms two little pouches over the milk-giving holes, and the little echidna when *very* tiny is put into this pouch, and keeps its head there while its body grows larger and sticks out beyond. In this way the Echidna can carry her child about with her, and she only turns it out to shift for itself when its prickles are hard and sharp.

* * * * *

You see, then, that though we began with the simplest known milk-giving animal, we are, in the Echidna, already fairly on our way to the curious pouched creatures of Australia, the "Marsupials," which, instead of a small fold, have indeed a large pouch of skin, into which they put their little ones when they are less than two inches long, and so imperfect that their legs are mere knobs, and they can do nothing more than hang on to the nipple with their round sucking mouths as if they had grown to it.

There the little ones hang day and night, and their mother
from time to time pumps milk into their mouth, while they
breathe by a peculiar arrangement of the windpipe, which
reaches up to the back of their nose. Then, as they grow,
the pouch stretches, and by-and-by they begin first to peep
out, and then to jump out and in, and feed on grass as well
as their mother's milk. For a long time they take refuge in
the pouch whenever there is any danger or they are tired,
and Professor Owen has suggested that this curious pouch
arrangement may be of great use in a country where water

Fig. 51.

Australian Marsupials.

Kangaroos; a flying Phalanger; and the Kaola or native Bear, with a young
one on its back.

is often so far to seek that the little ones could not travel to it unless the mother could carry them.

Now this race of pouched animals we find spreading all over a land where they had none of the higher four-footed animals to dispute the ground with them, for there are no ordinary land mammalia in Australia, except bats, which could fly thither; mice and rats, which could be carried on floating wood, and a fierce native dog, the Dingo, which was probably brought by the earliest native settlers long after the marsupials had spread and multiplied. And what is more, though we find the bones of marsupials of all sizes buried in the rocks of Australia, some of them as large as elephants,[111] showing that these creatures too had their time of greatness, we do not find those of ordinary mammalia.[112] It would seem, then, that for long ages the pouched animals had the field to themselves, and they made good use of it, filling all the different situations which in other parts of the world are filled by ordinary four-footed creatures.

On the plains, mountains, and red stony ridges are the long-legged Kangaroos every child knows so well in the Zoological Gardens. There they browse upon the grass and leaves as our cattle do in Europe, and some of them, such as the great gray Kangaroo,[113] grow to be as much as five feet high, and can make a good fight even against the fierce dingo dog, hugging him in their arms and ripping him up with the strong nail of the long middle toe of their hind foot, which answers in them to the hoofs of our cattle and deer. And yet they are peaceable enough unless attacked, as

---

111   Diprotodont.
112   The only exceptions to this are a tooth and a piece of a tusk of one of the ancient elephants, lately found in Australia, showing that a few straggling forms of mammalia probably reached that country in Tertiary times.
113   Macropus giganteus.

they lurk among the tall ferns and grass, and will far rather leap away than turn and attack an enemy. Others are much smaller, such as the Kangaroo Rats, which feed on roots and grasses, one of them, the Tufted-Tailed Kangaroo-Rat, [114] biting off tufts of grass and carrying them in his tail to make a soft nest to sleep in; while the Tree Kangaroos[115] of New Guinea live in the trees, feeding on the leaves and jumping from bough to bough.

All these, from their long hind legs and jumping movements, we should recognise at once; but the plump furry Wombat (see Fig. 52) looks more like an ordinary four-footed animal, as it wanders by night burrowing and gnawing the roots of plants. So too do the tree-climbing animals, the Kaola or tailless bear (Fig. 51), which often carries its young one on its back, and the beautiful Phalangers or "Australian Opossums," which live in hollow trees and come out on moonlight nights to feed upon the leaves, hanging from the boughs by their long prehensile tails. Yet all these animals have a pouch for their young, and while the long-tailed furry Phalangers play the part of the fruit-eating monkeys in a land where monkeys have probably never been, another group of them, the "Flying Phalangers" (Fig. 51), with a membrane stretching between their front and hind legs, represent the flying squirrels, and live at the very top of the gum trees, feeding on leaves and flowers, and taking flying leaps with their limbs outspread.

These are all vegetable-feeders; and they leave plenty of room for the little insect-feeders, the Myrmecobius, with its long bushy tail, and the Bandicoots or rabbit-rats, which

114   Hypsiprymnus penicillatus.
115   Dendrolagus.

feed partly on bulbs and roots, and more often on insects, grubs, and even small mice and vermin.

But where are the animal-eaters? Surely here, as in other parts of the world, some of the group have taken to feeding on their neighbours? There are very few carnivorous animals in Australia, and these are small, though fierce, and feed chiefly on rats and mice; yet the bones of huge marsupials, with long pointed teeth, found in the rocks, tell us that dangerous animals were once there before they were driven out, probably by the Dingo and savage man. And when we get to Tasmania, where no Dingos are found, there the flesh-eating marsupials still live, as fierce as any wolves and wild cats of Europe, and still they are pouch-bearers. Slim and elegant as the fierce and furry Tiger-wolf (Fig. 52) looks as he courses over the Tasmanian plains in search of prey, yet the mother carries her young in a pouch like the gentler wombat or the powerful kangaroo; and so does the mother of the Native Devil or Tiger-cat (Fig. 52), which is so fierce that even the natives are afraid of it when it turns at bay, and it will attack and devour large sheep, though it is only the size of a terrier dog.

We see, then, that the marsupials in a world of their own, cut off by the sea from the struggling world beyond, play all parts in life; and squirrels, monkeys, insect-eaters, gnawing animals, hoofed animals, and beasts of prey, all have their parallel among the pouch-bearers. But just because they are so isolated it becomes a curious question why, when we travel right across the wide Atlantic or Pacific to America, we find another set of pouched animals slightly different but belonging to the same group. How comes it that the clever little opossums of Guiana, Brazil, and Virginia (see Fig. 53, p. 194), which grasp the trees with the free nailless great toe of their hind feet and hang by their long tails, should be marsupials, carrying their little ones in pouches, when all their relations are thousands of miles away over the sea?

Fig. 52.

Tasmanian Marsupials.
The two to the left of the picture are Wombats;[116] the front right hand figure the Tasmanian Devil;[117] and the background figure the Tasmanian Wolf.[118]

Stop a moment, and let us go back to those times when the marsupials were living with the great flying reptiles in Europe and North America. These forms (see p. 178) were like the little myrmecobius now living in Australia, and at some period, we do not know exactly when, their descendants must have found their way to that part of the world, where they have since branched out into so many curious forms,

116   Phascolomys.
117   Dasyurus.
118   Thylacinus.

gnawing, leaping, running, and flying, and filling the place of ordinary quadrupeds. But they must also have lived on in the Northern Hemisphere and branched out into other forms; for much later, when tigers and other ferocious beasts had begun to prowl about in the forests of Europe and America, opossums were leaping in the trees, as we know by finding their bones in Suffolk, under Paris, and in North America. And so we see that when these opossums found their way down south to Brazil and Guiana, the simile that we used a little while ago (p. 130) probably became literally true, and the Australian and South American pouched animals are related to each other, not because they come one from the other, but because they both come from the same very ancient stock which once lived in Europe.

This would explain how these active, furry, little beings of all sizes, from that of a good-sized cat to a rat, come to be sporting among the leaves of the grand forests of Brazil or on the edges of the Virginian swamps, sleeping during the day in the hollow trees, and prowling by night over the plantations, and among the rice-fields feeding on fruit and seeds, worms and insects, and even on young birds and rats. On the ground they walk heavily, with flat feet, but in the trees they swing from bough to bough (see Fig. 53), the little ones curling their tails round that of their mother and clinging to her back as she goes. Some of these opossums have even lost the pouch, and put their little ones at once on to the thick fur of their back as soon as they come out of their snug nests in the tree-hollows. They seem to have a happy time of it, these merry tree-climbers, and know well how to swing out of danger, or to feign death if they cannot escape, so that "cute as a 'possum" is a common American proverb. One kind, living in the swamps of Guiana, feeds almost entirely on crabs, while another, called the Yapock, has webbed feet and dives under water, feeding on fish and other water-animals.

But here another question presents itself. How is it that these curious pouched animals have lived on in America as well as in Australia when they have been killed off in Europe and Asia? The answer to this is not far to seek if we remember that geology teaches us that there have been many changes of land and sea in past times, for the neck of land which joins South America to North America is very low and narrow, and a change of level of scarcely more than 2000 feet would break it up into islands; and as we know that such changes have taken place in past geological times, there is no doubt that once this neck was partially under the sea, and South America, like Australia, was a huge continental island, where the lower animals might struggle on and become settled, before the higher ones poured in to interfere with them.

* * * * *

Indeed, if the opossums did not teach us this history, we might learn it from another singularly old-fashioned race of animals; for in the same Brazilian forests in which our little opossums are sporting, the dreamy Sloth, with his long arms, short legs with the knees bent outwards, and long thick hair drooping over his eyes, is hanging back downwards from the boughs; while the strange Ant-bear is tearing open the ant-hills with his strong bent claws in the damp earth below, and licking up the insects with his long sticky tongue; and the Armadillo, whose back is covered with bony shields like the crocodile, issues out of his burrow at night to dig for worms or roots or buried animals. We may look all the world over and we shall not find another group so strange and old-fashioned as this one, nor even any creatures of their kind, except the ant-eaters of the Cape and the scaly Manises of Africa and India, which also live, as you will notice, upon continents which jut out into the water, and not on the great northern mass of land.

Fig. 53.

South American pouched animal, the Opossum;[119] and imperfect-toothed animals—Sloth,[120] Ant-bear,[121] and Armadillo.[122]

119　　Didelphis.
120　　Cholœpus.
121　　Mymecophaga.
122　　Dasypus.

In many ways these curious animals (*Edentata*) of South America and Africa are more singular, though not of so ancient a race, as the "pouch-bearers." Many of them, the American ant-bears and the African Pangolins, are quite toothless, and those which like the sloth have teeth, have very imperfect ones more like the teeth of reptiles than those of marsupials; again, their feet have the toes much joined together, and the sloths have only three toes on the hind feet and sometimes two only on the front, and the joints of their neck are irregular in number. Thus we see in them that variability of structure which always points to a low order of animals; and, moreover, the armadilloes are the only milk-giving animals which are covered with bony plates like reptiles.

Fig. 54.

African imperfect-toothed animals—Aard-Vark or Cape Ant-eater in the background, and scaly Manis or Pangolin in the foreground.

What, then, is the history of these old-fashioned animals? Much the same as that of the marsupials, so far as
we can read it; for at the same time that opossums were
living in Europe, strange animals, with imperfect rootless
teeth, and toes with immense claws, bent inwards like the
claws of the ant-eaters, were wandering over France and
Greece, where we now find their bones. Then, a little later
we find, on the shores of the Pacific in North America, other
huge imperfect-toothed creatures, which lived, died, and
were buried in the mud; and lastly, in South America, still
later, we find whole skeletons of gigantic sloth-like animals
the size of elephants,[123] which had not yet such long arms
as the Sloth of to-day, but walked on four feet upon the
ground and browsed upon the trees, while huge armadillo-
like creatures,[124] with solid bony shields covering their backs,
wandered in the vast forests and lived on animal food. Making
use of these facts, then, cannot we picture to ourselves how
these large unwieldy creatures, with their stiff bent claws
and their weakly teeth, which if once broken or lost could
not be replaced by a second set, were no match for the large
tigers, bears, and other beasts of prey which were roaming
over Europe and Asia; while those, on the contrary, which
found their way from North to South America, and were
cut off from the crowded world, just as the marsupials were,
might live on and fill the land with large creepers and burrowers. In the old world the same would probably happen
in Africa, where the sea certainly flowed at one time over
the low-lying desert of Sahara; and so the Cape Ant-eater
and the Pangolin, both so different from their American
relations, would keep their place in the world.

123   Megatherium.
124   Glyptodon.

This would explain how they gained a firm footing; but the next question is how they kept it, when jaguars and pumas began to roam over America, and lions and panthers over Africa? Now, if we inquire into the history of the Aard-Vark or great Cape ant-eater, which is in many ways much more like the American armadilloes,—for he has like them teeth in the back of his mouth, and walks flat-footed, though he has a thick skin and bristles instead of armour,—we find that he is a very timid animal, and lives almost entirely underground, only venturing out at night to scratch open the ant-hills with his strong claws, so that he may thrust his long sticky tongue into the ant-galleries to draw it back covered with food. Even then he never ventures far from his hole, so we can easily conjecture that it is by concealment that he has escaped destruction.

Still more would the Pangolins flourish, for though they are toothless and walk very clumsily, because their front feet are bent under so that they tread on the upper part, yet they have two means of protection. First, like the ant-eater, they live chiefly underground and come out at night; and secondly, their back is covered with sharp-edged scales, which grow from the skin as hairs do, and can be raised into a complete *cheval-de-frise* as they roll themselves up, or tuck their tail and head between their legs when they are attacked. Thus protected, the scaly ant-eaters not only flourish in Africa, but have even kept their ground in India, China, and Ceylon.

In America, on the other hand, we find that the armadilloes have gone strangely back to the bony armour of the reptiles or the ancient Labyrinthodonts, and have shields on their backs and heads formed of skin-plates exactly like those of the crocodile, so that the only delicate part of their body is the under side, which is kept close to the ground. When

we see how well they are protected, and also remember that
they are extremely quick burrowers and can get out of the
way of dangerous enemies, while they feed on vegetables,
insects, and dead creatures, we see why the plains and forests
of South America should abound in armadilloes of all sizes,
from the Great Armadillo, as large as a moderate-sized pig,
to the little Pichiciago, not larger than a rat.

It would be more difficult to understand how the great
hairy Ant-bear[125] (p. 194), with his twisted feet, united toes,
and toothless tube-like snout, has managed to live on in the
dense forests of South America, if we did not know that he is
immensely strong, and his sharp claws and the deadly hug of
his muscular arms are avoided even by large animals, while
the small American Ant-eaters[126] live chiefly in the trees,
feeding on bees, termites, and honey. A strange fellow is the
great ant-bear as he wanders at night slowly and heavily
along the river-banks, his long bushy tail sweeping behind
him and his head bent low; or, if it be a mother, she may be
carrying her little one clinging to her back, or pause to hold
it in her long arms as it sucks. Be this as it may, by-and-by
the ant-bear reaches a group of nests of termites (wrongly
called white ants), looming six feet high in the dark night; at
once the sharp claws are at work tearing the hill to pieces,
though they are so strongly built that men have to open them
with a crowbar, and as the alarmed termites rush out, the
long sticky tongue wanders among them and they are drawn
into the ant-bear's mouth by thousands. Yet the ant-bear
has his enemies, for it may be that in his night-walk he may
come across the fierce jaguar in search of prey.

Now, D'Azara, the great traveller, doubted the stories of

125   Myrmecophaga jubata.
126   Tamandua.

the natives when they said that the ant-bear could kill the jaguar, but Mr. Cumberland, who has lived much in South America and has himself killed the ant-bear, assures me that the animal is quite a match for such a wild beast. The muscles of his shoulder and arms are tremendous, the claws so hard and strong and sharp that when once stuck in they never lose their hold, and the ant-bear when attacked stands up and gives a death-hug so dreadful that the natives never dare to come to close quarters with him. Moreover, he is very difficult to kill. Mr. Cumberland, by the help of his dog and man, caught and disabled one of these creatures so as to tie his legs together and keep him stunned, but his skull was so hard that repeated blows with heavy quartz rock on his nose, the most vulnerable point, only succeeded in stunning him, and his skin was so tough that an ordinary small dagger-knife made no impression whatever. With all their efforts they could not put the poor animal to death till the following morning, when they could get a strong and sharp knife to butcher him. Such a creature as this need scarcely fear a jaguar or any beast of moderate size.

Such, however, is not the case with the dull-looking hairy forms which move among the tall cecropia trees above the ant-bear's head; for the sloths, though busy enough in the trees, would fare but badly if they were condemned to live upon the ground. The sloth is surely one of the most curious examples of how an animal may live and flourish by taking to a strange way of life. We have seen (p. 196) how his ancestors, the Megatheriums, walked upon the ground, while he himself was formerly pitied by all travellers because his arms are so long in comparison with his legs that if he wants to walk he has to drag himself along upon his elbows, and while the ankles of his hind feet are so twisted that he can only rest on the side of the foot. But then they forgot that he seldom or never descends to the ground, for the buds and leaves of the trees are his food, and they are so juicy that he does not need to come down to drink, and when he is in his natural place in the trees he is no longer helpless.

There, safe from prowling animals on the ground below, he hangs like a hammock from the bough. The long fingers of his hand (in some sloths two, in others three, in number) and the three toes of his twisted hind feet, all armed with long claws, seize the branch like grappling irons; while his long flexible neck, which in one kind of sloth has more joints than in other mammalia, enables him to look over his shoulder and take a wide survey around. In the daytime he sleeps with his back in the fork of a branch and his head bent forward on his chest, but as the sun goes down he rouses to life and feeds by stretching out those long arms to tear the leaves and twigs, which he stuffs into his mouth and chews with his few back teeth. He has no need to hurry or disturb himself, for his long thick hair protects him from insects; and from the very fact of his being fitted for a tree-life he is safe from other animals except snakes, and even they do not find him out easily, so like is his dull matted hair to the colour of the bark and moss. Even the young ones run very few risks, for they are not born till they are perfect, and then the baby sloth clings to its mother's hair, and goes with her wherever she travels, sucking till it is old enough to hang on to a bough and feed itself. So they live a completely tree-life, and sleepy as they seem, yet they can move quickly enough when they wish; and they often take advantage of a time when the wind is blowing so that the branches from tree to tree sway against each other, and by seizing the boughs as they touch, pass along and find new feeding-grounds.

We see, then, that while the duck-billed water-mole and the echidna have found a comparatively peaceful home in Australia, where the pouched animals have reigned as monarchs, and still hold their own in spite of the animals brought in by man; and while the opossums, by taking to a tree-life, revel in the forests of America: so the imperfect-toothed animals, an old and antiquated race of Life's children, still remain in a few scattered forms by reason of their power to adapt themselves to peculiar conditions of life. What they may have been in olden times we can scarcely guess; but one thing is certain, namely, that before such strangely different forms as the sloth, the ant-eater, the manis, and the armadillo could each have settled down and taken on their special protective armour and habits, many others must have tried, flourished awhile, and died out. When we look at the bones of the gigantic Ground-sloths or Megatheriums of olden times, which walked on four feet and are supposed to have lived by tearing the trees up by the roots and feeding on the branches, or when we examine the huge shield of the monster Glyptodon, and find that it had no movable bands between the plates such as enable the armadillo to burrow with ease, or in some kinds to roll up in a ball, we see that it is not always size and strength that win in the battle of life; but that the sloth of to-day has probably lived on because, in taking refuge in the trees, it has secured great advantages by those peculiarly long arms and twisted feet for which men used to pity it; while the ant-eaters and armadilloes in their underground homes, and the pangolins rolled up into prickly balls, show that passive resistance and retiring habits, especially if fortified by a thick skin, are sometimes quite as useful in the struggle for existence as fierce passions and aggressive weapons.

THE PIONEERS OF THE ARMY OF MILKGIVERS

CHAPTER IX.

FROM THE LOWER AND SMALLER MILK-GIVERS WHICH FIND SAFETY IN CONCEALMENT, TO THE INTELLIGENT APES AND MONKEYS.

HAVING now taken leave of the curious pouch-bearers and the strange primitive sloths and armadilloes, we find ourselves left to deal with an immense multitude of modern mammalia, which have spread in endless variety over the earth, and which may be divided into five great groups—the *Insectivora* or insect-eaters; the *Rodents* or gnawers; the climbing and fruit-eating Lemurs and Monkeys; the *Herbivora* or large vegetable-feeding animals; and the *Carnivora* or flesh-eaters.

All these groups are very distinct now, and we naturally turn back to ancient times to ask how they first started each upon their own road. But when we do this, we meet with a history so strange that it makes us long to open the great book of Nature still further, and by ransacking the crust of the earth in all countries to try and find the explanation, which will no doubt come some day to patient explorers. The history is this. We saw in the last chapter that in those far distant ages, when even reptiles were only beginning to spread and multiply by land and sea, and when, although birds probably existed, still they did not as yet leave any traces behind, small milk-giving and insect-eating animals, the *Microlestes* and *Dromatherium* (see p. 178), were already living upon the earth, and left their teeth and jaws in the ground.

Now, as ages passed on and the reptiles increased in strength, these little milk-giving animals evidently flourished, for though we have not yet discovered any of their bones in the rocks of the Chalk Period, yet as we find them both before and after that time, they must have lived on in some part of the world, the rich vegetation and abundant insect life affording them plenty of food. Meanwhile the huge reptiles, of kinds now long extinct, reigned over land and sea and air, and were in the height of their glory,—when suddenly there comes a blank and their history ends. When we look again, "a change has come o'er the spirit of the dream," and in the next period we find their bones no more. From that time we meet only with the four groups of lizards, snakes, tortoises, and crocodiles, which still survive; and the place of the swimming, flying, and walking reptiles is taken by four-footed and milk-giving animals.

Some of these were still marsupials like those that had gone before; others were of strange forms, distantly related to them; others were curious ancestral forms of our hyænas and bears, dogs and civets, horses and tapirs,[127] in which the characters which distinguish these groups were not so distinct as they are now, while others again were old forms of moles, hedgehogs, squirrels, bats, and lemurs. In what part of the world, then, had all these been growing up, that we come upon them so suddenly? Before the seas of the chalk only the small marsupials; after them, when the areas of land began to increase in extent, a whole army of milk-givers, so different from each other and so well adapted for their lives, that we even find among them such peculiar forms as whales, with their arms converted into paddles, and bats with their arms acting as wings.

What an idea this gives us of the immense period of time that must have elapsed while the chalk was forming, the reptiles becoming extinct, and the mammalia taking their place!

We have had a hint of this before, when we learned in *Life and her Children* how infinitely minute the shells are of which chalk is made, and what enormous thicknesses remain of the chalk-beds. And now we find these facts strengthened by the great changes which then took place in the animal world, for even if (as is likely) older forms of these large milk-givers existed in earlier times, and we have not yet found them, yet there are such great differences between

127   For a few of these forms see the picture-heading, p. 202.

whales, bats, dogs, and lemurs, that our imagination stands appalled at the time required to account for them.

Again, where are the traces of all the forms which must have existed between the little marsupials and this great army of four-footed beasts? At present no one can answer. Forty years ago we knew nothing even of those early marsupials, and people said there were no milk-giving animals until after the time when the chalk was formed. Now a few jaws have told us that milk-givers had been already in the world for ages; and it may be that before forty years more have passed, some child now reading these lines, and following in the footsteps of such patient explorers as Beckles and Gaudry, or the American naturalists Leidy, Cope, Marsh and others, who have such a grand field before them, may discover bones which will unravel the history of that crowd of mammalia which now seems to start up like Cadmus' army from the ground.

But for the present we can only begin with them as we find them immediately after the Chalk Period, and a strange motley group they appear. There, roaming among the palms, evergreens, screw-pines and tree-ferns, which flourished in Europe and North America in those warmer times, were beasts larger than oxen, with teeth partly like the tapir, partly like the bear, and feet like the elephant,[128] which may have been both animal and vegetable feeders. With them were true vegetarians, which could be called neither rhinoceroses, horses, nor tapirs, but had some likeness to each.[129] Others, half-pigs half-antelopes, were thick-skinned, but graceful and two-toed,[130] while a little fellow no bigger than a fox,[131]

128   Coryphodon.
129   Paleotherium and Anoplotherium.
130   Xiphodon.
131   Eohippus.

with five toes on his front feet and three behind, the ancestor
of our horses, grazed in the open plains. There too, moles,
hedgehogs, and dormice had already begun to make their
underground homes, and squirrels and lemurs sprang about
the trees of the forest, where bats roamed at night in search
of insects. Nor was this life without its dangers, for beasts
of prey, half-bears half-hyænas,[132] were there to feed upon
their neighbours, and with them a creature half-dog half-
civet,[133] with several other carnivorous animals with feeble
brains and partly marsupial characters,[134] and lastly a large
flat-footed dog-bear,[135] something between a dog, a cat, and
a bear, with a very small brain but plenty of teeth, represented the most primitive flesh-eating animal known to us.

    None of these forms were of the same species as those now
living, and many of them, as we see, had characters which
we now find in two or three different animals; showing that
they had not yet specialised the various weapons of attack
and defence, and the difference of limbs and teeth which
now distinguish their descendants. So that, for example,
though there were fierce animals of prey, none had yet the
formidable teeth of the tiger nor the muscular strength of the
lion, neither had the vegetarians the fleetness of the horse,
the horns of the deer, nor the large brain of the elephant.

    This had all to come with time, and from that day to this
their descendants have been spreading over the earth. Some,
large and powerful, have conquered by strength; some, by
superior intelligence, have learned to herd together and
protect each other in the battle of life; some have gone back
to the water and imitated the fish in their ocean home; and

132   Hyænarctos.
133   Cynodon.
134   Hyænodon and others.
135   Arctocyon.

others, smaller and feebler, have lived on by means of their insignificance, their rapid multiplication, and their power of hiding, and feeding on prey too minute to attract their more powerful neighbours.

Among all these there are hundreds of different forms, branching out here and there, crossing each other's path and often jostling on the way; while during the long period between our first knowledge of them and now, they have been driven or have travelled from one country to another, from the northern to the southern hemisphere, or from the Old to the New World, till in many cases it is impossible to say what routes they have taken.

How, then, shall we get a glimpse of the nature of these large groups? Shall we take the moles and hedgehogs as the lowest, and the monkeys as the highest, and then travel in a straight line through the forms between? Scarcely, I think, for it is very doubtful whether the lemur and the dormouse may not be able to boast of ancestors as ancient as the moles, while the elephant and the dog are surely as intelligent and far nobler animals than the monkey. No! we must make up our minds at once that the different branches have grown side by side to much the same height, so that our genealogical tree, if it were possible to make one, would, like a real tree, be a mass of entangled twigs, some of which would, indeed, be less aspiring than others, yet on the whole we could scarcely say that one reached nearer to the sky than another. What perfection they have each obtained in their own line is quite another question, and one which we are able to trace out.

* * * * *

Thus, for example, the gnawing animals or *Rodents*, and the insect-eaters or *Insectivores*, are undoubtedly the lowest types next to the sloths and armadilloes, the insect-eaters especially having very primitive skeletons and small brains. Yet we shall find that we pass very naturally from them to the intelligent monkeys, while, on the other hand, the vegetable-feeders and flesh-eaters go off upon quite a different line of their own.

Let us, then, begin with these two lowly groups, the *Rodents* and *Insectivores*, and see how they have conquered their humble place in the world. One thing is clear, that they do not hold it by strength or audacity, for taken as a whole they are small and weak animals; the giants among rodents, the Capybaras of South America, where all lower kinds of animals thrive, are only as large as good-sized pigs, and the smallest, the "Pocket-mice" of North America, are not bigger than large locusts; while the insect-eaters have nothing larger than the "Tenrecs" or soft-bristled hedgehogs of Madagascar, about the size of a tailless cat; and the rest of the group vary from two to eight inches all over the world. Moreover, they are as a rule timid, and though some of them fight fiercely among themselves, yet they scamper away and hide at the least alarm, and generally choose the twilight or the dark night for their feeding time.

Stroll out some fine summer's evening, when the sun has set and the moon has not yet risen, and as you wander in the fields and woods with eye and ear open, you will scarcely have gone far before you will be aware that there is plenty of stir going on. Some active little field-mouse will cross your path in her eager search for grain and seeds to lay up for her winter store, or you may startle a hare in the long grass and watch her run across the field, or see her sit upright on her haunches surveying the quiet night-world. Or, if you pass over a common, the number of little white tips glancing in the twilight from under the furze bushes will tell you that the rabbits have not yet disappeared into their burrows; while as you enter the wood the sharp little eyes of the squirrel will peep down upon you from the beech trees, as she watches over her little ones in their comfortable nest in the branches.

All these are *Rodents*, and you may know them by their four long chisel-like front teeth (see B, Fig. 55), which have a large gap on each side, between them and the grinding teeth behind. These chisel teeth have not bony roots like the teeth of most animals, but rest in a deep socket, and continue growing during the whole of the animal's life; and they have a hard coat of enamel in front, so that as the tooth wears away behind, this enamel stands out and forms a sharp cutting edge, and there is perhaps no tool more efficient for gnawing a root, a nutshell, or the solid wood of a tree, than the tooth of a beaver or rat.

Fig. 55.

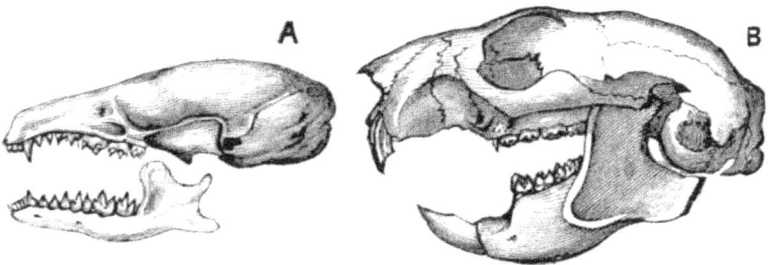

A, Skull of an insect-eating animal (*Insectivore*), showing the numerous pointed teeth. B, Skull of a gnawing animal (*Rodent*) showing the large chisel teeth in front, and the gap between these and the hind teeth.

But these animals have another and quite a different set of companions, as you will learn if you are lucky enough, by looking carefully along the hedge, to startle a little shrew in its quest for worms, or to catch a hedgehog shuffling along at a sharp trot after his nightly meal of beetles, slugs, and snails; nay, you may even, if it be early summer, come across a mole, or find two fighting fiercely together for possession of the only thing they come to the upper world to fetch—a wife.

These creatures have not the long front chisels of the hare or the shrew; on the contrary, their mouth is small, and crowded with a number of fine pointed teeth (see A, Fig. 55), of which even the back ones have sharp cusps or points, well fitted for crushing insects. For these are *Insectivora* or insect-eaters; and while the rodents are gnawing at roots and leaves and nuts, these devourers of small fry mingle with them very amicably; while both groups only ask that the night-owl may not see them in their evening wanderings, nor the weasel and his bloodthirsty tribe attack them in their homes.

For, ever since they began the race of life, long long ago, these two very different orders of animals have been trying to feed without risk, and to keep out of the way of flesh-eating birds and larger creatures. And so it has come to pass that, though the rodents are mostly plant-eaters, while their associates are insect-eaters, yet, as both are trying to conceal themselves, and get their food by stealth, they have acquired curiously similar external forms, weapons, and habits of life, with the one exception of their teeth and the manner of eating their food.

Even in our English meadows a casual observer might easily mistake the little *insect-eating* shrew, with its soft velvety coat and bare paws (Fig. 56), for a near relation of the *gnawing* Harvest-mouse nibbling the grass tips just above its head (Fig. 57); though a nearer inspection of the shrew's long snout, small ears, and sharp teeth, would show the difference. And as to their way of life, the Field-shrew and the larger

Field-mouse live like two brothers of the same race. They both make burrows in the banks, though the field-mouse digs the deeper hole, and they both line their home with dry grass to bring up their little ones. And when the winter comes they both retreat into their homes; the shrew to sleep away the dark days, and the mouse to wake from time to time to feed upon his store. Only their food is quite different, and when they come out in the twilight of the summer's evening, the mouse is on the look out for acorns, nuts, grains, and roots, which it gnaws off with its sharp chisels, while the shrew is chasing worms and insects, or cracking tiny snails with its pointed teeth.

Then if you lie and watch quietly by the bank of a river, there you may see the Water-rat or Vole (not the land-rat which sometimes hunts for prey in the water) diving under with a splash to gnaw the roots of the duckweed or the stems of the green flags, and coming up to sit on the bank, and hold them in his paws as he eats them; while not far off a pretty little Water-shrew, this time too small and different to be mistaken for his companion, is swimming along with his hind feet, the air bubbles covering his velvety back with silvery lustre as he chases water-shrimps, or feeds on fish-spawn or young frogs. Both these animals live in streams and rivers, and bring up their young in holes in the bank, where they can jump into the water if the weasel attacks them, or the common snake pokes his head too near their home.

These are perhaps the chief examples we shall find in England of insect-eaters and gnawers living near together and following the same kind of life; but if we look over the world it is most curious how many parallels we can draw between them, showing how the same dangers have led to the same defences.

Fig. 56.

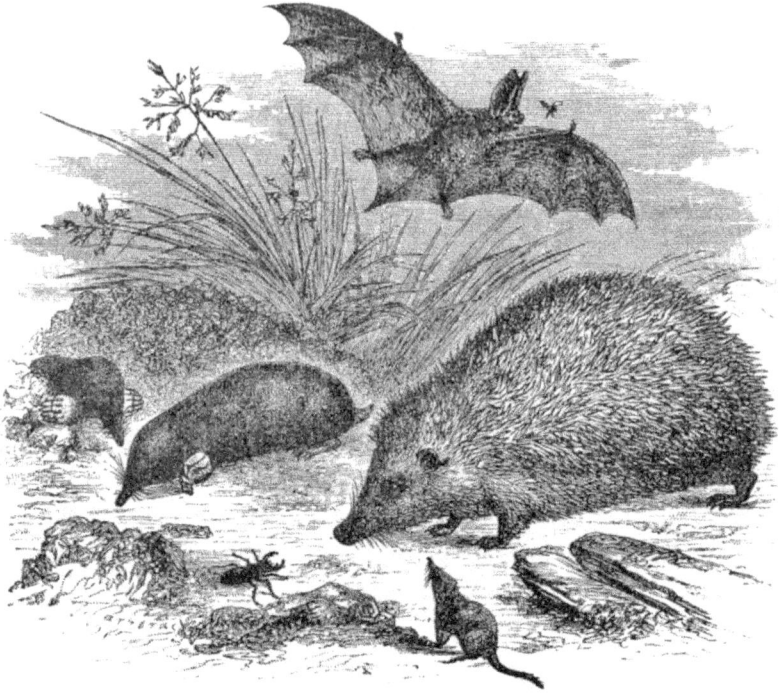

A group of Insect-eaters.
Common Shrew, Hedgehog, Mole, Bat.

Look among the *insect-eaters* at our Hedgehog (Fig. 56), so weak and shuffling in his movements that he would have been cleared out of the world long ago but for the sharp elastic spines which grow upon his back in the place of hair. There he goes trotting along under the hedges in the twilight, cracking the horny skins of beetles, or sucking eggs, or devouring worms, slugs and mice when he can get them, without a thought of fear. For he can roll himself up in an instant if danger be near, and his sharp spines will keep off even dogs and foxes, unless they can catch him unawares, and bite him underneath in his soft throat. Nay, he can actually master a poisonous snake, and use it for food, not suffering even from the adder's fangs when they pierce his tender nose.

Fig. 57.

A group of Rodents.
Harvest-mouse, Porcupine, Mole-rat.

It is curious to see how quickly he can roll himself up by drawing together the strong band of muscle which passes along the sides of his body from head to tail, sending out bands of muscle to feet, head, and legs. When he contracts this band his limbs are all drawn in, and the spiny back forms a kind of prickly bag all round them, even his tender snout being safely hidden. Nor are his spines merely sharp—they are as elastic as the hair of which they are modifications; and the hedgehog can drop safely from a height when he is in his ball-shape, falling on the spines, which bend and straighten again as though made of whalebone. So he lives under hedges and in ditches till the winter comes, when he settles down in a nest of moss and leaves in a hedgebank or a hollow tree, and sleeps the cold weather away. And when the spring comes he takes a wife, who brings up her little ones in the nest of moss and leaves under the hedgerow, watching over them as long as their spines are soft.

And now where shall we look among the *rodents* for a creature to match the hedgehog among insect-eaters? Surely to the "fretful Porcupines," which feed on all kinds of vegetable food in Southern Europe, Africa, Asia, and America, protecting themselves by the formidable array of spines which they can raise at will. Even the European porcupine, which is about two feet long and the weakest of his tribe, is better protected than is generally believed. It is true that his long black and white ringed spines only cover the hinder part of his body, but the hair of his head and neck hides a number of short spines which can give very sharp pricks; and though he is a timid night-loving animal, hiding by day in burrows and holes of the rocks, yet when attacked he jerks himself up against his enemy, so that the long spines wound very severely. And when we come to the Tree-porcupines, which hang by their tails from the palm trees in Mexico and Brazil, we find that their short stout spines are a very efficient defence both against birds of prey and the deadly coils of the boa constrictor and other large snakes; while the Western porcupine and the almost tailless Canada porcupine, which climb trees and strip off their bark and buds, have a clothing of such dangerous weapons that pumas and wolves have been known to die of inflammation from the wounds.

The porcupine among the rodents, then, like the hedgehog among insect-eaters, has adopted prickles as a defence. But there are many soft-haired creatures living upon the ground in both families which have no protection but concealment, and we find them both gaining it by burrowing into the ground. Among the *insect-eaters* the Mole is the most successful digger, and as he works his tortuous way through the ground in search of worms and grubs, it is scarcely possible to imagine a miner more usefully equipped for his work. His skeleton, it is true, is, on the whole, more primitive and roughly finished than that of higher animals, his ear is almost closed, and his eye though bright is deeply hidden; but the parts specially necessary to him are most wonderfully fitted for the work they have to do.

His broad shovel-like front paws (see Fig. 56), with their five strong claws, set each in a long groove at the tip of the last finger-joint, are powerful tools for shovelling away the earth, as he turns them outwards and pushes with them as if he were swimming; and they are carried on strong, short, and broad front legs, fixed to collar-bones and a shoulder blade of unusual strength, while the breastbone is so formed as to throw the legs forward and bring them on a level with his nose when he is burrowing. This nose, too, has its part to play, for it is long and slender, with a small bone at the tip, which helps him in pushing his way forwards while his hind feet are planted flat and firm on the ground behind, while it also serves to pick out the grubs, worms, and beetles from their narrow holes.

Here, then, we have the very best of miners, who has secured food and safety far from the busy world above, and spends his time hunting for grubs and earth-worms in the dark earth below. He is a most voracious animal, and makes the ground above him heave and swell as he toils through it eager for prey, pushing up every now and then with his nose the loose earth he has excavated, thus marking the line of his route by molehills.

But when he builds his home and fortress where he takes his long winter's sleep, and hides from weasels and pole-cats, he takes care to throw no loose rubbish above; on the contrary, he presses the earth together so as to make the walls of his chamber firm and hard, and carries out from it a number of passages, by any of which he can reach his home in safety when he is pursued too closely.

Thus by his cleverness in burrowing, and the useful tools which he carries upon his body, the mole has managed to find safe feeding-ground and shelter, when no doubt many of his relations living above ground have been killed off. Even underground he has his enemies, for the Weasel, the Stoat, and the Badger find him good eating, while if he meets one of his own brothers in a narrow passage they will fight till one is killed and eaten; yet though fierce he is also tender-hearted, for mole-catchers say that when a mother-mole is caught in a trap the father may sometimes be found dead by her side.

And now if we turn to the *rodents* for rivals to the mole, we are almost confounded by the multitude of creatures which have found safety in burrowing. Not only have we the rabbit-warrens, by which the sandy soil of our commons is riddled in every direction with holes, leading to burrows where the mother lies snugly hidden with her five or six naked little ones in a bed of her own fur; but we have the extensive burrows of the little, long-legged, leaping, gnawing Jerboas of Africa, which are so like the Jumping Shrews among insect-eaters. Then again there are the underground cities of the South American Viscachas and Chinchillas, and the extensive subterranean settlements of the Lemmings,—those curious rodents, which from time to time start off in vast swarms across Norway, over mountain and valley, through flood and fen, over rivers and plains, preyed upon by eagles and hawks, foxes and weasels, on their way, but never stopping or swerving in their course till they reach the sea, into which they plunge and drown themselves. Again, every inhabitant of Switzerland knows the Marmot and the burrows he forms, scratching up the earth with his hind feet and patting it together with his front paws and his broad nose; while every American child has heard of the hillocks thrown up by the "Prairie Dogs,"[136] which undermine whole plains in the far west with their underground cities, where the burrowing owl shares their home with them, and the rattlesnake steals their young.

Fig. 58.

The Pyrenean Desman,[137] an insect-eating water animal.

136   Cynomys.
137   Myogale Pyrenaica.

But all these come out upon the land and use their burrows chiefly for homes and nurseries. We can match the mole better than this among *rodents*, for in Eastern Europe, India, and Africa, there are blind creatures called Mole-rats[138] (see Fig. 57), with broad flat heads, small eyes hidden in their fur, short tails, and feet with sharp claws, which live almost entirely underground, burrowing subterranean galleries in the sandy plains in search of roots, as the mole does for worms; while the Pouched Rats[139] of North America also live in burrows, throwing up hills just like mole-hills, and gnawing roots and buried seeds, which they carry in their large cheek-pouches, to store up in their underground chamber for winter food.

Nevertheless, the rodents can scarcely compete with the mole as burrowers, and it is not till we come to the water-animals that *they* begin to have the best of it. True, the insect-eaters have the Water-shrew and the curious West African Shrew,[140] with its broad tail; while the Desman[141] of Russia and the Pyrenees (see Fig. 58), with his dense furry coat, his broad tail, and his webbed feet, is quite a match for the gnawing Musk-rat or Musquash of North America, for they both live in fortresses on the river-banks, to which hidden passages are well contrived to elude pursuit; and while the desman, with his curious movable snout, pokes about in the Russian or Pyrenean streams after leeches, water-snails and insects, the musquash in America gnaws off the roots and stems of water-plants.

138   Spalacidæ.
139   Geomyidæ.
140   Potamogale.
141   Myogale.

Fig. 59.

The Beaver,[142] a gnawing water-animal.

But the insect-eaters have no water-animal to match the Beaver in sagacity, judgment, or engineering. For here we have a creature not much larger than a good-sized cat, cutting down trees, dragging logs six feet long to the water's edge, and building with them the most elaborate log-houses and water-dams. With hind feet webbed up to the claws, and his broad tail as a rudder, the beaver has so much swimming power that his fore legs are free to carry and place the wood, while his broad orange-coloured teeth, as sharp as chisels, which grow as fast as he wears them away, are his cutting instruments. With them he gnaws a deep notch in the trunk of a larch or pine or willow, as deep as he dares without fear of its falling, and then going round to the other side, begins work

142   Castor fiber.

there till the trunk is severed and falls heavily on the side of the deep notch, and therefore away from himself. Then, after stripping off the bark and gnawing the trunk into pieces about six feet long, he uses his fore-paws and his teeth to drag them into position to build his dam. The lighter branches he uses to make his oven-shaped lodge, laying them down in basket-work shape, plastering them with mud, grass, and moss, and lining the chambers with wood-fibre, and dry grass; and the logs he piles up to form dams, lest at any time the stream should flow away and leave the entrances to his home dry. These dams are very skilfully and cunningly formed. He always makes the deep notch in the trunk on the side near the water, so that the tree in falling comes as near as possible to the stream; then he does not always clear away all the branches, but he and his companions place the logs with these lying *down* the stream, so that they act as supports to resist the current and prevent the dam being washed away. Thus they make a broad foundation, sometimes as much as six feet wide, and upon this they pile logs and stones and mud till they have made a barrier often ten feet high and more than a hundred feet long.

In this way they clear the woods just round their stream, as if a whole gang of wood-cutters had been there at work; and as the dams check back the water and form broad meres, there are soon swamps on all sides, where peat moss grows and "beaver-meadows" are formed.

Here the beavers live in companies, each in his own chamber with his wife and family, though underground passages often lead from one to the other, and when water-plants and soft bark are scarce, they will often travel some way inland to feed on fruits and grain. But if among the community any are lazy or will not take partners, they are driven out, to find a refuge in holes of the river-banks, where they sulk alone.

In Western Europe, indeed, where they have been so much persecuted, most of the beavers live alone in holes, though communities are still left in parts of Germany, Scandinavia, and Siberia. But in North America they still carry on their true communal life, and those who visit their wonderful settlements will not be surprised to learn that they possess the largest brain for their size of any of the gnawing animals.

Indeed, they would have no rival among rodents if it were not for the clever sagacious rats, and these have probably sharpened their wits by living so long in contact with man, for they are burrowers chiefly in human dwellings, granaries, stables, mines, ships, and every available dwelling-place where they can rob and plunder, and outwit even man himself by working their way into his stores, and acting together in carrying away his goods.

So the insect-eaters and rodents hold their own both by land and water, penetrating, in the forms of bats and mice even to Australia, though the rodents are most widely spread, for except two very rare animals[143] in the West Indian Islands, there are no Insectivora except bats in South America. The bats, however, remind us that both these groups have also found homes above the ground and in the trees. There the *rodents* have the lovely little Squirrels, which, with their brown red backs, white waistcoats, and graceful bushy tails, scamper up the trees of our English woods. It is very tempting to dwell upon the squirrel, with his little wife, to whom he remains faithful all his life, his beautiful round nest, in which his young are so carefully reared, and his pretty ways as he sits upright gnawing beechnuts or acorns, holding them in his tiny hands. He has made good use of his opportunities, being almost as widely spread as the rat, for there are squirrels of some kind all

143   Soledon and one of the Shrews.

over the world, wherever there are forests, except in Australia. Several of them in the East and North America have folds of skin at the side of the body, which, when tightly stretched, by extending the four limbs, enable them to take flying leaps from tree to tree (see Fig. 60). Even without flying, however, the squirrel is so nimble that he manages well to escape his enemies, except some of the birds of prey and the fierce tree-marten and wild cat; and as in cold countries he sleeps soundly in snug holes of a tree till the leaves grow again to give him shelter, he is not often detected even by these.

Fig. 60.

On the tree, the Taguan[144] or flying squirrel, a rodent;
Flying below, the Colugo,[145] an insectivorous animal.

144   Pteromys Petaurista.
145   Galeopithecus volans.

Nevertheless, in tree-life and in the air it is the turn of the *insect-eaters* to claim the advantage. It is true that the insect-eating Bangsrings,[146] which scamper up the trees in Sumatra and South-East Asia, and were long mistaken for squirrels, are a small family and not of much importance; but what shall we say to the Bats, the only true flying milk-givers? Or what, again, to that curious animal the Colugo or Flying Lemur of the Malay Islands, which belongs to the insect-eaters, and yet has some points like marsupials, some like fruit-bats, and some like the true lemurs? This strange creature, which seems like the remnant of some branch-line from very ancient times, climbs the tree like a squirrel by means of its claws, and then spreading out its limbs displays a broad membrane (see Fig. 60) stretching not only along its sides but across its tail, and from the front of the arms to the neck as in bats, and so sails down from one tree to another. The mother, which Mr. Wallace examined, nurses the little one on her breast just as the lemurs do, while large folds of her skin protects the small, bald, naked little creature, something after the manner of an imperfect pouch. Lastly, while they sometimes feed on insects, the chief diet of these colugos is fruit, like the lemurs, to which group they were once supposed to belong.

* * * * *

But of all modified insect-eaters the most extraordinary are the Bats, which are so different from all the others that they have been placed in a distinct order[147] of their own. Imagine a little creature about three inches long, with a body something like a shrew, large ears, a protruding snout, and plenty of sharp teeth (see Figs. 61 and 62). Let it have

146   Tupaia.
147   Chiroptera.

Fig. 61.

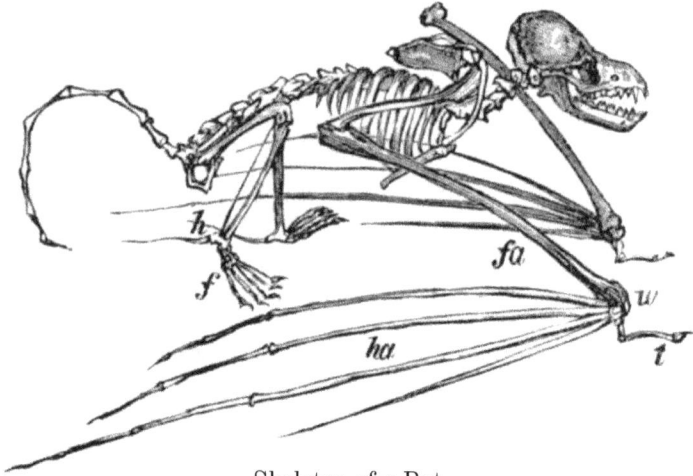

Skeleton of a Bat.
(Lettered to compare with bird's skeleton, p. 126).
fa, fore arm; *w*, wrist; *t*, thumb; *ha*, hand; *h*, heel; *f*, foot.

a breast bone projecting more than in most milk-givers, and covered with a large mass of muscle as in birds, fitted to move the wings, but having nipples to suckle its young. Let it have large shoulder-blades and collar-bones, a strong upper arm, a very long lower arm (*fa*, Fig. 61), and four immensely long fingers to its hand (*ha*), and a short clawed thumb (*t*). Let its hind legs be short and weak, with a long spur behind the heel (*h*) of its five-toed feet, and finally let the skin of its body grow on over the arms and long fingers, filling in the space between the elbows and the neck in front, and stretching away behind, over the legs down to the ankle, and on behind the legs, so as to enclose the tail. This skin growing from the back above, and the under part of the body below, will enclose the bones of the arms, hands, and legs, like a kite with calico stretched on both sides (see Fig. 56, p. 212), and when the long fingers are outspread and the legs opened, no limbs will be seen, but only a small body and head, with an immense expanse of skinny wing, from which the short clawed thumbs and the four toes of the feet stick out before and behind.

Now this creature is no longer like the flying squirrels or the colugo, which can only take floating leaps; for though like them it has only a membrane stretching out from its body, yet this has become a long flexible wing, formed on a widely outstretched arm and abnormally long hand, and moved by powerful muscles like the wings of birds or insects. It is essentially fitted for flitting through the air in search of prey, while it makes but little use of the running power which it possesses in common with all other insect-eaters. If you see a bat moving along the ground, you will acknowledge at once that it is a true quadruped, yet, by its awkward gait as it shuffles along on its clawed thumb and toes, you will judge that it is not an earth-loving animal. Watch it at night on the wing and it is quite another creature; then it will flit about in and out of cracks and crevices, under the eaves, round the haystacks, or among the trees, and never once strike its wings against anything, though it has been proved that it does not trust chiefly to its bead-like eyes to guide it.

Bats have been blinded, their ears stopped with wool, and their noses with sponge dipped in camphor; and yet, without sight, hearing, or smell, they steered quite successfully between outstretched threads or tree-branches, or found their way into a hole in the roof. In truth, as they have become fitted to navigate the air, they seem also to have become sensitive to its currents. Their wings are abundantly supplied with nerves and blood-vessels, and have little rough points all over the surface; their ears have generally a second ear-lobe or leaf within the outer one, and those which have not this have leaves of skin or membrane round their nose. With all these they seem to feel the slightest difference in the air, so as to detect at once whether they are in the open, or whether any resisting object is near them.

Fig. 62.

A Bat walking.

Now it is clear that a creature of this kind, able to chase insects in the air, even in the darkest night, can secure much food that the running insect-eaters can never reach. When the little common English bat, the Pipistrelle, awakes from his day's sleep, which he has been taking, head downwards, hanging by his feet in some old tree or under the roof of a barn, he finds the gnats and flies abroad, and begins his chase in the twilight—up and down, from side to side he flits, and his wide-open mouth takes in insects at every turn. And by-and-by, as the dark nights come on, the Long-eared Bats begin gradually to stir from their clusters in the barns and old buildings, and, unfolding their wings so as to display their ears as long as their bodies, commit sad havoc among the night-moths. All night long their shrill squeak may be heard, but before day dawns they are away again, and may be found hanging in dense masses by their hind legs to the timbers of some old church belfry, or in caves, or even under the roofs of houses, where they find an entrance by some hole, and go in by hundreds to hang from the rafters.

Many accounts are given in American writers of the thousands of bats collected in the caverns which abound in the Western States, while in the Egyptian catacombs they hang in myriads. For of all things a bat dreads the light when beasts of prey are abroad, and next to that he fears any position near the ground where weasels, wild cats, or other flesh-eating animals may seize him in his sleep. Nay, the smaller bats live in constant fear of the larger ones, for they feed upon one another with evident relish.

Yet in spite of dangers the bat family, aided by its power of flight, has spread all over the world, from the Arctic Circle to the Equator, east, west, north, and south. In cold countries they hang by their feet in the winter, or sometimes by their clawed thumbs, and sleep in dark recesses, scarcely breathing till the warm weather and the insects return; but in warm countries they are active all the year, sleeping by day and feeding by night.

In England and North America they are content chiefly with insect food, but in South America the Vampires, among the leaf-nosed bats, fasten on to large animals and suck their blood. Mr. Darwin had his servant's horse bitten and disabled for two days by a vampire in Chili; while Mr. Wallace, when on the Amazon River, was himself twice bitten, once upon the great toe, and once on the tip of his nose while asleep! A bat is a grotesque-looking animal at best; but some of these leaf-nosed bats are simply hideous, with their wide-open mouth, sharp teeth, and the skinny leaves sticking up round their nose.

How different are the gentle-looking fruit-eating bats of the Tropics, which seem to belong to quite a different branch of the family. Their fox-like and intelligent faces are a pleasure to look at, reminding one of the lemurs, and harmonising beautifully with their quiet peaceful life among the fig-trees, guavas, mango-trees, and plantains of the East. There they hang in dense masses from the tall silk-cotton trees till night comes on, and then take wing as soon as the sun is set, and hooking themselves by one thumb to the fruit-trees, hold the fruit in the other as they feed.

Thus we have a wide range of habits in bats, from the insect-eaters to the blood-sucking vampires on one hand, and the gentle fruit-bats on the other.

But one virtue the most bloodthirsty and the most gentle have in common, and that is maternal love. As soon as the little ones are born they cling to their mother's breast, and she often folds over them the skin which covers her tail, so as to form a kind of pouch, so that wherever she flies they go with her, and are carefully tended and suckled by her till they can take up the chase for themselves.

And now we have followed out the Rodents and Insectivora in their various lines. Both lowly groups, of simple structure and with comparatively feeble brains, they have chiefly escaped destruction from higher forms by means of their nocturnal and burrowing habits or arboreal lives, and

Fig. 63.

Fruit-bats[148] hanging from the ledges of a cave in the Mauritius.

the marvellous rapidity with which they breed, combined with their power of sleeping without food during the winter in all cold countries. Nevertheless, though they are often strangely alike in outward form, they differ in many remarkable respects. The insect-eaters now existing are chiefly a few straggling forms of a once widely-spread group; while the rodents, on the contrary, are still a very numerous and varied family, spread all over the earth, and boasting of such intelligent forms as the squirrel, the beaver, and the rat. But

148   Pteropus vulgaris.

here their advantages appear to end, while the insectivora point onwards not only to the bats, the only flying milk-givers, but also through the colugo to the lemurs, and thus onwards to the monkeys. It may be, and indeed probably is true, that the colugo started off from some very early type, more nearly related to the pouch-bearers than the present insect-eaters are; while the monkeys, again, branched off long ago on another line quite separate from the modern lemurs. But if the tiny shrew wished, like many little people, to boast of distinguished connections, he might with justice suggest that somewhere among his primitive ancestors one would probably be found whose descendants had risen far higher in the world than himself.

* * * * *

It may perhaps seem strange to many readers that instead of leaving the apes and monkeys to the last, as standing at the head of the animal kingdom, we should bring them in now, directly after such lowly creatures as hedgehogs and mice, bats and beavers. It must, however, be repeatedly borne in mind that we are not following a direct line upwards, but a family tree, which branches in all directions; and though the gap between monkeys and insectivora may be great, yet they have many more points in common than the monkeys have with any of the vegetable-feeders or carnivorous animals, and probably we should find these links even more marked if it were not that we know so very little of the early his-tory of Monkeys. The reason of this probably is that they live and die in woods, where any remains of their bodies not eaten by other animals decay and crumble to dust, so that we have only here and there a few skeletons to tell any tale of their ancestors. And so it comes to pass that when we first meet with the great army of milk-givers (see p. 202), lemurs, and soon after true monkeys, existed, with thumbs on their hands and grasping great toes on their feet.

In those times, when the climate of Europe and North America was warm and genial, they spread far and wide with the other animals over Germany, England, and the United States, where forests of palms, fig-trees, and evergreens afforded them a congenial home. But as soon as these began to fail and the climate of the northern countries became cold and cheerless, we find the monkey-kingdom growing narrower and narrower, till in our own day, while the flesh-feeders range from the Arctic Circle to the Equator, and the vegetarians have their reindeers travelling over ice and snow on the one hand, and their hippopotamuses and giraffes wandering under the burning sun of Africa on the other, the tender monkeys, which shiver in cold and damp and are constant victims to consumption, have shrunk back into the Tropics, where there is abundance of fruit and vegetation for their food. It is true a few kinds still linger in Japan, and one[149] on the sunny Rock of Gibraltar, while one or two wander up the mountains of Tibet into the regions of frost and snow; but, on the whole, monkeys are essentially inhabitants of warm countries, where the trees are perpetually covered with leaves and fruit, as in the luxuriant forests of South Asia and Tropical Africa in the Old World, and Tropical America in the New.

Though they have but a narrow kingdom, however, there can be no doubt that they make the most of it, and have managed to develop shrewdness and a sense of fun and frolic which would be quite unaccountable if it were not for one peculiarity which they possess. This peculiarity is the grasping power of their hands and feet, which has caused

149   The Magot, Macacusinuus.

them to become such active nimble creatures, swinging, leaping, and running quickly along the boughs of the tangled forests in which they live.

Yet the monkeys do not stand alone in this grasping power, for we have seen that the opossums have hind-thumbs among the pouch-bearers, while among the rodents the little dormouse has a nailless grasping toe-thumb on his hind feet. So that here already we have some clue to possible descendants of poor relations of the monkeys down in the lower forms of life; and when we remember that the colugo (see p. 223) is related on the one hand to marsupials and insect-eaters, while on the other it leans towards the lemurs, and through them to the monkeys, we begin to suspect that somewhere low down in all these groups we might find ourselves among a family party from which all the different branches have sprung; just as we found the birds, reptiles, and milk-givers starting in past ages among the amphibia.

It must, however, be very long ago since the monkeys scrambled to the top of this family tree, for even the Lemurs,—which are not true monkeys, but a lower type with an irregular number of teeth like the insect-eaters, hairy hands and fox-like faces, without any change of expression,— have well-developed thumbs and toe-thumbs, with nails on hands and feet, and they have besides that free movement of the arm and wrist which gives at once an advantage to the *Quadrumana*[150] or four-handed animals.

These lemurs are a gentle and loving race of creatures, which run on all fours like cats, and have none of the mis-

---

150   Naturalists now class monkeys under the order "Primates" (or highest forms), together with man, and they have given up the term Quadrumana, or four-handed, because, although the feet grasp like hands, they are *true feet*. Nevertheless, this term is very useful; and, if properly understood, expresses the grasping power of the four feet characteristic of the group.

chievous half-reasoning pranks of monkeys. They must have crept down long long ago from the great battlefield of Europe and Asia, and taken refuge in the forests of South Africa and India, and especially in the Island of Madagascar, where they were sheltered from the attacks of larger and fiercer animals. They are splendid climbers, with very sensitive tips to their fingers, which are often of different lengths, and many of them have eyes with pupils which expand and contract like those of a cat, enabling them to see well by day and night, while a quick sense of hearing warns them of any danger near.

In India, indeed, their relations the "Lories" are most of them slow-moving night-loving animals, while in South Africa the "Galagos" sleep all day in a nest of leaves, and are only active at night, crying to each other as they leap from bough to bough, seizing the beetles and moths in their little hands. It was probably from such night-wanderers as these that the general name of "lemurs" or "ghost-like" animals was given to the group, for the *true* lemurs, which live in Madagascar,—their special home, where they have few enemies,—may be seen by day running along the branches, snatching the fruit, sucking birds' eggs, and even feeding on the young birds themselves, for they have plenty of crushing teeth, as well as incisors for clipping the leaves. Sometimes they sit in companies, huddled together, wrapping their soft furry tails round each other's necks, for they are chilly creatures, and even in that warm country their thick tails, which are quite useless for clinging, seem to be a comfort to them. More often they are running and jumping, especially in the evening time, the mothers carrying their naked little ones nestled in the fur of their stomach, or, when they are older, on their backs; and whether slow or quick, day-lovers

or night-hunters, these happy thoughtless little beings flourish in the quiet island home they have found, cut off from
the struggling world beyond.

Fig. 64.

The Aye-Aye and a Lemur in the forests of Madagascar.

And among them at night, when the soft clear moonlight shines down on the thick forests in the interior of the
island, comes a small ghost-like animal, the "Aye-Aye,"
with wide-staring eyes, furry body, and long bony jointed
fingers. He utters a plaintive cry as he creeps from bough
to bough, stripping the bark off the trees with his strong
chisel-like teeth to find some worm-eaten hole into which he
thrusts his skinny fourth finger to pick out a grub, and then

moistens his meal by drawing the same long finger rapidly through some watery crevice, and then through his lips for drink. This strange creature too is a kind of lemur, so far as he can be classed at all, with his gnawing teeth, his hind feet like a monkey's, his large spoon-shaped ears, and his uneven fingered hands, with strong curved claws. At any rate he belongs to no other group, but tells us once more the old story of creatures in isolated countries putting on strange shapes suited to extreme habits of life.

Now between these gentle, but low-brained and dreamy lemurs, and the active, intelligent, mischievous monkeys, there is a great gap. The creatures most like them are the little Marmosets of South America, which run like squirrels among the forest trees of Brazil, feeding on bananas, spiders, and grasshoppers, and making their nests in the topmost boughs. But these marmosets are true monkeys, with expressive faces, and the peculiar wide-spread nostrils which we find in all the monkeys of the New World. For it is to South America, that land of the less advanced forms of life, that we must look for the lower kind of quadrumana, with side-opening nostrils,[151] thumbs which move in a line with the fingers of the hand, and not nearly so much across the palm as in the higher apes, and thirty-six teeth in their mouth instead of thirty-two,[152] as in man and in the Old World monkeys.

None of these American monkeys ever become so man-like as the Apes of Africa and Asia, but in many ways they bring monkey-life in the trees to greater perfection, in the dense forests of Brazil and Paraguay, and even as far north as Guatemala. The lumbering heavy Gorilla of

151   Platyrrhine monkeys, from *Platus* broad, *rhines* nostrils.
152   Except the marmosets, which have a peculiar dentition of their own.

Fig. 65.

A Woolly Monkey and child (Lagothryx Humboldtii),
showing grasping tail. (Proc. Zool. Soc.)

Africa, though higher in the scale, is a cumbersome fellow
compared to the nimble little thumbless Spider monkeys of
the Amazons, which hang by their bare tipped tails to the
branches and to each other, chattering away like a troop
of children as they gather the bananas and other fruits, or
catch insects and young birds, or fly screaming with fear
from the stealthy puma or the fierce eagle. With the trees
for their kingdom, their tail for a fifth hand, and the warm

sun to cheer and invigorate them, these spider-monkeys and their quieter friends the Capucine monkeys (often seen on London organs), and the Woolly monkeys (Fig. 65), lead a pleasant life enough, till misfortune or old age overtakes them. Their friends the Howler monkeys, which also have grasping tails, seek the deep recesses of the forest and creep quietly from tree to tree until night comes, when hundreds of them at once will make the woods re-echo with their deep howling cry, which they produce by a special voice-organ in their throat; and with them come out the little Owl monkeys, which sleep by day in the hollows of the trees. These, with the various kinds of Saki monkeys, which cannot cling by their tails, but have fairly good brains and quick intelligence, make up the monkey population of America.

Here, then, we have a whole group of quickwitted tree-monkeys, which, from their structure, we know must have started long ago on a line of their own, wandering down into South America, where they had but few enemies except the boas and pumas and birds of prey, till man came to kill and eat them. And if we wonder how they have gained their quick mischievous intelligence in those quiet pathless forests, we must remember that though a grasping hand and foot seem at first sight of very little importance, yet by means of them the monkey moves rapidly from place to place, swinging, leaping, running, and climbing along the boughs, which are its paths from tree to tree. And since rapid change of any kind makes the eye quick of sight, the ear acute, and the brain active and alive to take in new impressions, it is no wonder that the monkey mind has become alert and ready during the ages that these animals have been chasing and cheating and outwitting each other, or tenderly rearing their young ones among the dangers of the forest.

And now if we turn back to the Old World, it is not so much the smaller active tree-monkeys that interest us, for they live much the same life as their American cousins, although they differ from them in never having grasping tails, in having thirty-two teeth like man, in the openings of their nostrils which turn downwards[153] like our own, and in having either cheek-pouches to stow away their food, or stomachs with three compartments like animals that chew the cud, so that they can keep a store within. But in spite of these differences they appear outwardly much the same as the American monkeys; they leap and jump among the trees, and it is not till we come to the Baboons and the tail-less man-like apes, that we find ourselves studying quite another kind of life.

Imagine an undulating country of corn-fields and rough vegetation in Abyssinia, or southwards towards the Cape, with long ranges of rocky hills rising up behind, and preci-pices leading to the narrow defiles of the mountains, and then picture to yourself, descending from those mountains, a troop of two hundred or more large hairy monkeys, with short tails growing from between bare seat-pads, dog-like faces and something of a dog's shape, as they gallop clumsily along with all four feet flat upon the ground. These are the African Baboons, and they form a goodly company, the chiefs marching first, grand old elders with stout hairy manes to protect them when fighting. These come cautiously, peer-ing over the precipices, and climbing up rocks and stones to survey the country round before allowing the troop to advance; and behind them follow the young males, and the

153   Catarrhine monkeys; *kata* downward, *rhines* nostrils.

mothers with their children on their backs, shambling down till they reach the fertile grounds, where sentinels are set to watch for danger, while the multitude feed, filling their cheek-pouches and even storing the corn under their armpits. Then when all are satisfied, if no alarm has been given they wander slowly back, resting by the way to chew their food or drink at some mountain stream, but never leaving the company till they are safe back under the rocky ledges of the steep hillside, where they make their home.[154]

For these baboons, unlike other monkeys, live in hilly rocky places, and not in forests, and therefore they are in much more danger from wild beasts, especially the leopard, so that they rarely venture abroad except in company, and lead an extremely gregarious life. Yet though they run on all fours, and look less human than most monkeys, even the lowest baboon, the Mandrill (easily known by the coloured swellings on its cheeks and hind quarters), which has many points in its skeleton like four-footed animals, has true thumbs on its hands and toe-thumbs on its feet, and uses them to lift up stones to search for scorpions and other insects; while the mother baboons dandle their little ones, or give them a box on the ear when troublesome, in true human fashion.

Moreover, they have developed great intelligence in their social life, and the youngsters are soon taught to keep silence when danger is near, to follow their leader, and to obey the sign of command; while, in their turn, the leaders will defend the weak and feeble of the troop, as in the well-known case of the brave old baboon who came down alone in the face of the dogs to fetch away a little one only six months old, which had been left behind crying for help.

154   See Parkyns' *Life in Abyssinia*.

Still, notwithstanding their cleverness and courage, these baboons, with their long hind legs and dog-like faces, running on all fours, travelling in troops, and feeding in the cornfields and meadows, remind us more of four-footed animals than any other of the monkey tribe, and we must turn again to dense forests and tangled jungles to find those large and tailless apes which have risen highest in monkey life.

If we go back in imagination to those days when the wild beasts of the forests, the strong elephants and rhinoceroses, the fierce tigers, lions, and leopards, had not yet been persecuted by man, but roamed in great numbers over the whole tropical and temperate world, we can easily imagine that a set of animals which could climb along the tops of the lofty trees in impenetrable forests would have a great advantage, even though elephants, rhinoceroses, and buffaloes were crashing through the underwood below, and the fierce leopard was on the watch for them when they ventured to descend. With their tree-loving life, the monkeys would have every chance of escape, climbing along the topmost boughs with wonderful rapidity, to find refuge in gloomy recesses where they might bring up their young in safety. And as they grew in strength and intelligence, gradually retiring to the thickly wooded part of Southern Asia and tropical Africa, they might even succeed in driving out their opponents, as the Gorilla is said to have driven the elephant from the Gaboon country, because he interfered with the trees which he makes his special home.

So we must go to such tangled virgin forests as those of Sumatra, Borneo, and Malacca, to find the long-armed tailless Gibbons,[155] which once wandered over Europe, but now roam no further than Southern Asia, where they swing

155   Hylobates, or walker in the woods.

themselves along from branch to branch by means of their lengthy arms, which are so out of proportion to their legs that when they stand upright they can touch the ground with their knuckles. These gibbons are gentle creatures, with not too much brain, but wonderfully elegant and agile, which is more than can be said for the intelligent Orangutan[156] or Mias which wanders in the same forest. He has shorter arms, only reaching to the ankle, and he climbs half upright from tree-top to tree-top, grasping the boughs and swaying slowly onwards, or holds on by his toe-thumbs while he stretches up to the more slender branches to gather the fruit and young buds.

A strange object he looks, a great red, hairy, man-like creature, between four and five feet high, thrusting his huge black face from out of the dense foliage as he devours the Durian and Mangosteen fruits, seated comfortably in a fork of the tree, and then if disturbed he is off far more quickly than you would suppose possible for such a heavy creature, running, climbing, and creeping half upright till he is lost in the forest. He rarely comes down, except to shamble across some open space from one wood to another, or to drink in the river, where the natives say the crocodile attacks him, but he beats him and carries off the victory; while in the trees his only enemy is the python, which tries to encircle him in its coils. Nor does he often wander in company, for Mr. Wallace tells us that he never saw a father and mother orangutan together, though either of them may be seen with the young ones. He seems to lead, on the whole, a solitary life, and when the sun goes down retires into a nest of leaves low down in one of the trees, and sleeps till it is broad daylight and the dew is dried off the leaves.

156   Malay: *Orang* man, *utan* forest.

But, though the orangutan is both strong and cunning, he is not nearly so human as the intelligent and docile Chimpanzee, which shares with the fierce Gorilla the dense forests of palms, amomas, and gigantic tropical trees of Africa, where the grass and brush grow fifteen feet or more high, and the native man scarcely dares to venture for fear of the man-like apes. In these endless African forests there is quite a population of these wild creatures; bald-headed apes which build bowers in the trees; the Soko, a kind of gorilla, which loves to steal the native children, and always defends himself by biting off the fingers or paws of his enemy; the true chimpanzee, so human in its affection and its fun when it is caught and tamed; and the fierce gorilla, between five and six feet high, which rules as master in Western Africa near the equator.

Though each of these tailless apes has its own advantages, yet the gorilla is, on the whole, most advanced and nearest to man in structure. But his legs are still too short and thick, and his arms long, reaching to his knee; and the large projections on the back of his neck bones prevent him throwing his head well back, so that he stoops like a hunchback, while his feet are twisted so that he treads on the outside and not on the sole. His eye-teeth are huge, his eyes deeply sunken, his jaws heavy and strong, but his brain is not one-half the size of that of the lowest races of men, and though it has foldings very like those of the human brain, these are larger and less complex. When he walks it is not upright but on all fours, resting the knuckles of his hand on the ground; but when he is in his natural home—the trees—then his long strong arms and broad naked palmed hands grasp the boughs with immense power, and pull his heavy body upwards as he climbs hand over hand, his twisted

Fig. 66.

The Gorilla at home.

toe-thumbed feet clutching the branches below far better than a straight foot could do.

And so he lives with his wife and family in the thick solitary parts of the West African forests, feeding only on fruits and leaves, so that his stomach becomes large and heavy with the amount of food necessary to nourish him. He is more sociable than the orangutan, for several will travel together, but he asks for no shelter beyond the trees and the nest of leaves, which is his home and the cradle of his young ones, nor does he seem to attack other animals except in self-defence, and then his gigantic strength and his formidable teeth are his chief weapons, and woe betide the creature that comes within his grasp.

It is strange to picture to ourselves these huge apes, living in the depths of lonely forests and looking like human savages to those who can catch a glimpse of them, so that the ancient Carthaginians landing on the shores took them for "wild men" and "hairy women." We know very little of their daily life, for they are seldom seen except by those who hunt them, and who have but little chance of watching their habits. But all that we do know teaches us that in their rough way they have developed into strangely man-like though savage creatures, while at the same time they are so brutal and so limited in their intelligence that we cannot but look upon them as degenerate animals, equal neither in beauty, strength, discernment, nor in any of the nobler qualities, to the faithful dog, the courageous lion, or the half-reasoning elephant.

CHAPTER X.

THE LARGE MILK-GIVERS WHICH HAVE CONQUERED THE WORLD BY STRENGTH AND INTELLIGENCE.

IF we now glance back in imagination over the almost end-less variety of creatures which we have met with since we started with the fish, we must acknowledge that even if there were no other kinds than those we have already mentioned, the world would be very full of different living beings, and that to succeed in the struggle for life in the midst of such a multitude, new forms must be endowed with great strength or armed with specially effective weapons.

Such animals, however, we know were already in the field, for we saw at the beginning of the last chapter that, together with the small rodents, insect-eaters, and lemurs, there were two groups of much larger animals, first the *Herbivora* or grass-feeders, including the hoofed animals (*Ungulata*) and the elephants; and secondly, their great enemies the *Carnivora* or flesh-feeders.

Now these two groups, on account of their size, strength, and agility, have spread very widely over the earth, especially the grass-feeders, for there is no part of the world which has not some vegetable-feeding animal in it, if only a few green shoots grow there. It is true the Rodents take some part of this green food, but then they are small and insignificant compared to the large Rhinoceroses, Elephants, Hippopotamuses, Oxen, Antelopes, Goats, Pigs and Sheep, which roam over wide spaces, and are even less restricted than the flesh-eating animals, for they live in the open air or the thick jungle, never in caves and holes, and their young ones are born wherever they may happen to be, and in a few hours run by their mother's side, so that young and old wander together wherever food and shelter is to be found.

And so we shall see that these vegetable-feeders have filled every spot where they could possibly find a footing. In the regions of snow and ice the reindeer in Europe, and the elk and musk-sheep in America, rake the snow to uncover their scanty food, while the burning deserts of North Africa and East Asia have bred their camels and wild asses, and those of South Africa their quaggas. On the prairies of America the bison, and on the plains of Asia the wild cattle, feed in herds of thousands, while the zebra courses over the African hills. If we look to the tops of mountains, to dangerous crags where the merest tufts of grass are to be found, there we meet with the goats and sheep in India and Asia, the chamois and ibex in Europe, the big-horn sheep in the Rocky Mountains of America; or if we turn to the dense forests and tropical jungles, there we find the giraffes in Africa, the elephants, rhinoceroses, buffaloes, antelopes, and wild boars in Africa and India, some feeding on the branches of the trees, some grazing on the grasses and lower brushwood, and some digging up roots and underground food. Only the rivers remain, and here too, in Africa, the hippopotamus has taken possession, feeding on the water plants and wallowing on the muddy banks.

In this way every available spot is used by one herbivorous animal or another, and if we could only trace out their pedigree we should be surprised to find how wonderfully each one has become fitted for the special work it has to do. But three things they all require and have, though they may arrive at them in different ways. The *first* of these is a long face and freely moving under jaw, with large useful grinding teeth to work up and chew the vegetable food; the *second*, a capacious stomach to hold and digest green meat enough to nourish such bulky bodies; and the *third*, good defensive weapons to protect themselves against each other, and against wild beasts. Weapons of attack they do not need, except for fighting among themselves; for being grass-feeders they do not attack other creatures, and this is one of the great differences between them and the flesh-feeding or *carnivorous* animals.

We need not look far to see these three chief characters

of the vegetable-feeders in active work. Look at any horse as he grazes in the meadow, and see how his under jaw works from side to side as soon as he has a good mouthful. A peep into his mouth will show that he is using broad flat back teeth to grind the grass to pulp (see Fig. 67), and he will go on eating all day without overfilling the large stomach which lies within his barrel-shaped body. And as to his defences, if he is vicious, he will soon show that his front teeth are good weapons, while his hoofs will deal an ugly blow.

Then turn to the cow, quietly chewing the cud by his side; you will find that she has no upper front teeth, but only a hardened gum, upon which her under teeth bite as she crops the grass; but she too has broad flat teeth behind, while within she has a stomach with four compartments, and when she has filled one of these full of half-chewed grass, she lies down, and with a slight hiccough returns a ball of food to her mouth to be leisurely ground down. It is not difficult to see that to animals, such as wild cattle, antelopes, goats, and sheep, which often have to go far to seek their food, an arrangement of this kind, by which they may store provender in a larder for quiet enjoyment by-and-by, must be a great advantage. But the cow cannot defend herself with her teeth since she has no upper ones in front; in their stead she has strong horns which are quite as dangerous, so that an angry bull is an enemy not pleasant to meet.

Lastly, there is another fierce vegetable-feeding animal almost as dangerous as a bull, though we no longer come across him in England; for the Wild Boar, as he still flourishes in the forests of Germany, can inflict very ugly wounds with his lower eye-teeth which grow out and project over his upper lip, forming large tusks.

So we see that while the vegetable-feeding animals have

three characters in common, namely, large flat grinders, a capacious stomach, and defensive weapons, their defences, on the other hand, may be of three different kinds, and they may depend upon horns, hoofs, or teeth for protection.

Now in the beginning, when we first meet with the milk-givers, these defences were not so complete in any of the vegetable-feeders as they are now. Of the elephants alone it may perhaps be said that they had large and formidable ancestors.[157] As to the rest, the huge hippopotamus and sharp-tusked boar were only represented by small animals;[158] and even later, when the hogs branched off in a line of their own, they had at first only ordinary teeth, which did not grow out as tusks.

So, too, the fierce horned rhinoceros had as an ancestor a hornless tapir-like creature,[159] and the graceful hoofed horse a little creature no larger than a fox, with five separate toes on his feet.[160] Lastly, all the horned animals which chew the cud,—oxen, buffaloes, antelopes, and deer,—were nowhere to be seen, and in their place were only some small elegant creatures without horns.[161]

It is only at a later period when the flesh-feeding animals grew strong and dangerous, and the vegetable-feeders had to struggle for their lives, that we begin to find the remains of hogs and hippopotamuses with tusks, rhinoceroses with nose-bones, and fleet horses which could take to their heels, or bite and kick their enemy to death; of stags with antlers,

---

157   The Dinocerata of the Middle Eocene of America. These gigantic extinct animals, with tusks and horns, but very small brains, are believed by Professor Marsh to have connected the two groups the elephants and the hoofed animals among the early milk-givers.
158   Anoplotherium; for this form and others, see p. 245.
159   Paleotherium.
160   Eohippus.
161   Xiphodon.

ever increasing in size; and of bulls and buffaloes, goats and antelopes, with true horns. For not only by this time were they persecuted by the flesh-feeders, but they themselves were becoming very numerous, and it was the strongest only that could secure feeding-grounds or carry off wives.

* * * * *

It is very curious to see the different ways in which the three chief lines of vegetable-feeders secured these advantages to themselves. First, there were the hogs and hippopotamuses. The hogs did not grow to any enormous size, but their thick skins were a great protection to them, and their eye-teeth became their defence, growing out from the lower, and sometimes from both jaws into huge tusks; while their broad, round, flexible snouts served them to turn up the ground, and so get at roots and underground fruits such as other grass-feeding animals could not find; though at the same time they did not despise snakes or toads, and have become omnivorous animals. And so they have spread nearly all over the world; in Europe and Asia as wild hogs, and their wives the sows; one peculiar form, the Babirusa, being found only in Celebes; in Africa as large Wart-hogs, some as big as donkeys, with two pair of strong tusks curling out of the mouth; while in South America the family is represented by the small Peccaries, which travel about in herds, and have no tusks to show; but which, nevertheless, are bold and fearless, for they have *within* their lips short lancet-shaped tusks, which inflict fearful wounds. Only in North America, north of Texas, no *wild* creature of the hog family now lives, though in ancient times there were plenty of them.

Fig. 67.

The Babirusa; the double-tusked hog of Celebes.

Meanwhile the warmth-loving hippopotamuses, the hog's nearest relations, with huge grinding teeth behind, sharp front teeth, and tusks within their lips, took to a water-life in the Old World.[162] When we look at their immensely powerful bodies, and their short stout legs with four strong hoof-covered toes, and learn how rapidly they can gallop on land, and how furiously they charge an enemy in the water, snapping their great jaws which will kill a large animal at

162   See picture heading.

one crunch, we do not wonder that they can hold their own, especially as they always live in herds. Yet large and powerful as they are, they have not spread far over the earth, for though in past ages the hippopotamus swam in the river Thames, and grazed and left his bones in the ground upon which London streets now stand, yet after a time they crept down to warm Africa, where they may now be seen lazily basking on the surface of the Nile or of the river Zambesi by day, and making tracks by night into the swamps and jungle to feed on the coarse rank grass. They are well fitted for their life, for their thick naked skin, with pores which give out a fatty oil, keeps them from chill in the water; their eyes are set well back on their heads, so that as they float deep they can still look around, and the slits of their nose, and the openings of their ears, can both be closed and made water-tight when they dive, while their slow breathing enables them to remain a long while under water.

The second line was that of the rhinoceroses, tapirs, and horses, or the uneven-toed animals which have one or three toes on the hind feet. They took to very different means of defence. The Tapirs,[163] large, heavy, and with enormously tough hides, seem to depend chiefly upon their great strength for defence. Starting in warm times in the Old World, they have wandered in their day nearly all over the globe, dying out in later times, till now one kind is left solitary in Sumatra and Malacca, and the remainder have found their way down to South America, where they tear the branches from the trees with their short movable snouts, and feed peaceably at night unless attacked, when they make a furious rush at their enemy and conquer by sheer force.

The rhinoceros, the tapir's nearest relation, is even bet-

163   See picture heading.

ter defended; his skin is so thick and hard that in the Indian rhinoceros it actually forms a kind of jointed armour; his skull is wonderfully strong, and his nose is supported by thick bones, on the top of which are one or two solid horns, which are formed by a modification of the hairs of the skin growing matted together.

And now notice, just as we saw that the horned cow has no front upper teeth, so too the rhinoceros, though his horn is of quite a different kind, has in some cases lost his front teeth, which he does not need, since he rushes with his horn at his enemy instead of biting. Like the hippopotamus, the rhinoceros once wandered all over Europe and Asia, and when the great cold came on, the woolly species which roamed far north was often caught in the frost and snow

Fig. 68.

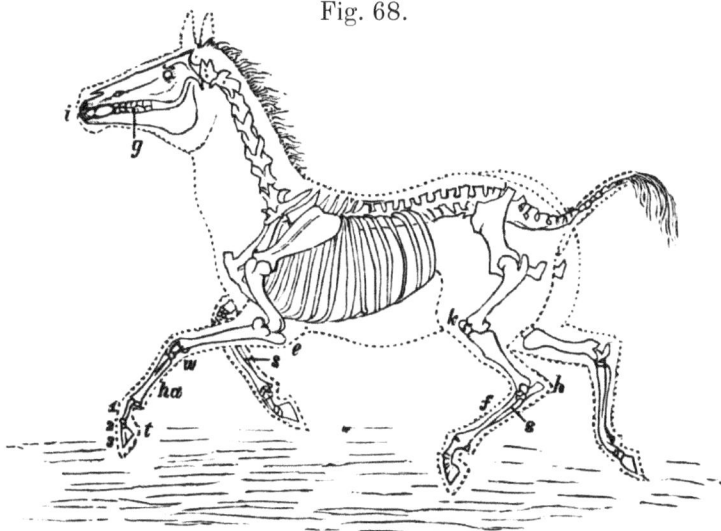

Skeleton of a Wild Ass.

i, incisor teeth; *g*, grinding-teeth, with the gap between the two sets as in all large grass-feeders; *k*, knee; *h*, heel; *f*, foot; *t*, middle toe of three joints carrying the hoof; *s*, splint, or remains of one of the two lost toes; *e*, elbow; *w*, wrist; *h*, hand-bone; 1, 2, 3, joints of the middle toe.

of Northern Asia, where his fleshy body has been found
preserved in the ice. Now he too has taken refuge in the
warm parts of Asia and Africa, where he either grazes on
the plains or plucks the leaves from the trees in the jungle
with the fleshy flap of his upper lip.

But of all the animals of this three-toed group the Horse
has the most interesting history, because we can read it
most perfectly. The only certainly original wild animals
of the horse tribe now living are the Zebras, Quaggas,
and Asses of Asia and Africa; yet strange to say, it was in
America that this tribe began, for there we find that tiny
pony[164] not bigger than a fox, with four horn-covered toes
to his front feet (and traces of a fifth) and three toes on his
hind ones. Then, as ages went on, we meet with forms, still
in America, first with four toes on the front foot, and then
with only three toes on all the feet, and a splint in place of
the fourth on the front ones. In the next period they have
travelled into Europe, and there, as well as in America, we
find larger animals with only three toes of about equal size.
One more step, and we find the middle toe large and long,
and covered with a strong hoof, while the two small ones
are lifted off the ground. Lastly, in the next forms the two
side toes became mere splints; and soon after, in America
and in Europe, well-built animals with true horse's hoofs
abounded, the one large hoof covering the strong and broad
middle toe. For what we call a horse's knee is really his wrist,
and just below it we can still find under the skin, those two
small splints (*sw*) running down the bone of the hand, while
the long middle finger or toe, with its three joints (1, 2, 3),
forms what we call the foot. It is by these small splints the

164   See p. 205, and picture heading, p. 202.

horse still reveals to us that he belongs to the three-toed animals.[165]

Now while these changes in the toes were going on, the space between the front teeth and eye-teeth gradually increased, till we arrive at the large gap now seen in the horse and ass (see Fig. 67). The chief bone of the fore arm (radius) increased in size, and the other bone (ulna) became joined to it, and the same in the hind leg. The brain increased in size mainly in the front part, and the body grew much larger, improving in form and build, till the long, slender, flexible legs became the perfection of running and galloping limbs such as we find in the zebra of to-day, poised upon a strong jointed toe, with its last joint broadened into a firm pad, and covered with a thick nail—the hoof. We have only to compare the well-proportioned leg of a horse with the thick, strong, clumsy leg of an elephant, to see, on the one hand, what a shapely and beautiful limb it has become; while, on the other hand, if we put it by the side of a giraffe's leg, we must acknowledge at once that it is a far stronger and more serviceable limb than if it had gone to the other extreme. There can be no doubt that when the horse arrived at this

165   The genealogy of the horse is so important, that it may be well to give a table of the seven principal stages, though transitions are known even between these.

Period.	In America.	Front Toes.	Hind Toes.	No. of Teeth.	In Europe.
7. { Recent . . . . — } { and } { Upper Pliocene . Equus }		1 / 2 splints	1 / 2 splints	40	Equus. } Equus. }
6. Upper Pliocene . . Pliohippus		1 / 2 splints	1 / 2 splints	42	—
5. Lower Pliocene . . Protohippus		1 large / 2 small	1 large / 2 small	44	Hipparion.
4. Upper Miocene . . Miohippus		3	3	44	Anchitherium.
3. Lower Miocene . . Mesohippus		3 / 1 splint	3	44	—
2. Upper Eocene . . Orohippus		4	3	44	—
1. Lower Eocene . . Eohippus		4 / 1 splint	3	44	—

point of the strong single hoof and well-shaped body, he had a wide range over the world, both Old and New; but curiously enough, while in Asia and Africa the tribe branched out into many forms, such as asses, quaggas and zebras, in America it died out, so that till we found the fossil-forms,[166] it was thought that no horses had ever been there till they were brought by the Spaniards.

Meanwhile, in the Old World, they must have led as free and joyous a life as those horses do now which have run wild in Tartary and America, galloping, frolicking, feeding, and neighing to each other with delight, as they roamed over the wide plains in troops of thousands, for solitary wanderers they would soon have fallen a prey to wolves or jaguars; and if the mothers wished to protect their foals they had to learn to follow one leader and act together in time of danger.

> "A thousand horse, the wild, the free,
> Like waves that follow o'er the sea,
> Headed by one black mighty steed
> Who seemed the patriarch of his breed,"

they grew accustomed, as generations passed on, to unite against their common foes, placing the mares and their foals in the centre when attacked, while the fathers met the enemy with hoofs and teeth. And so they became intelligent and tractable even in their wild state, to those of their own kind, and laid the foundation of those noble qualities of which man now reaps the benefit.

But the horses were not the only group which combined in this way for protection. The third great line of hoofed animals, those which have "cloven" feet of two toes, and which "chew the cud," have learnt many a lesson of vigilance, fidelity, and affection, by their social habits. Everyone has

166　　See table, above.

read of the herds of antelopes or deer, where the sentinels stand faithfully watching while their companions feed, and stamp or whistle when danger is near; while in the herds of wild cattle, not only will the mothers keep a watchful look-out for danger, but the bulls will join to protect the young ones at the risk of their own lives. Mr. Allen relates how, in America, a young bison, which had strayed from the troop and was followed by wolves, was surrounded by a number of old bulls, who, facing about, warily conducted him across the plain till he was safely among the dense mass of buffaloes, which the wolves dared not attack.

Now these "ruminant" animals, with complicated stomachs and the power of feeding at long intervals, have spread far and wide over the earth under many different forms, and while some are still very numerous, others are now rare, or almost destroyed.

Take, for example, the Camel, the true "child of the desert." There are no wild camels left now, so long has man conquered and tamed this useful beast of burden. But in past ages vast numbers of camel-like forms lived in North America, which found their way on the one hand to the south, where the Llamas, Alpacas, and Guanacos now feed on the mountains of Peru and Chili, while on the other they travelled over Northern Asia to the deserts of Africa and Arabia, and there became those curious desert-animals which the Arabs used and still use as their beasts of burden. A strange old fellow is the camel, with his two-toed hairy feet, with only nail-hoofs upon them, and his hard pads on his thighs and legs, on which he rests when he lies or kneels. His curious fleshy hump, which is single in the true camel or dromedary and double in the Bactrian camel, serves him as a special provision of fat, and it dwindles when he is short of food,

Fig. 69.

The true Camel (Camelus Dromedarius).

recovering its size and firmness when he is full-fed again; and he is the only cud-chewing animal which has kept his front teeth and defends himself with them, having no horns.

Still more strange in some ways are the giraffes,[167] of which we know very little, except that large forms like them once wandered in Europe.[168] For they, with only the same number of bones as other animals, have these so lengthened out that, as they wander in the tropical forests, their slender legs raise them above all other animals, and their long neck, which nevertheless has only seven joints like all the milk-givers, enables them to reach the high trees, so as to strip off the leaves with their ribbon-like tongues.

---

167   Cameleopardalis.
168   See heading of chapter.

Fig. 70.

The Red-deer with branching antlers.[169]—(*AFTER RIDINGER*).

But we should want much space to discuss such curious forms as these, and we need not go further than the ordinary deer of our parks to read a strange history of how life has gradually armed her children. The giraffe with his long neck to feed, and his wide straggling legs to fly swiftly from danger, has only short hairy covered knobs on his forehead for horns. But the stag, who is obliged to fight, especially when he wishes to secure his wives, has antlers so branched and so heavy that it is a wonder that his neck can carry them.

Now it is in the autumn that the stags fight and struggle together to secure the leadership of the does, and it is then that their antlers are finest and strongest, and they remain so during the whole winter. But when the early spring comes, the bone of the antlers dries up near the head, where there

169   Compare this with the Deer with the one-spiked antler in the picture heading.

is a little ridge round it, and soon they fall off, a skin forms over the place, and new ones begin to grow. Then as the little knobs push forward and increase, how lovely they are, for the skin covered with soft hair is all over them, carrying the network of blood-vessels which secrete the bone within. So fast do they grow that antlers weighing seventy-two pounds will be complete in ten weeks, and when they are finished, the "velvet," as this soft skin is called, dries up, and they rip it off against a tree, leaving the bare bone.

Thus equipped, the stag is a match for the world, and he knows it; his bearing is proud and haughty, and instead of flying from danger he will turn round and fight fiercely when attacked. And now comes the curious part of his history. In the different stags of the world we see all kinds of antlers, from one single spike like a stiletto in some American stags, to the superb antlers of the Red-deer, some of which have as many as sixty-six spikes. But when the red-deer begins to grow his antlers, he does not get this splendid tree in the first year, he has only a single spike; this falls off, and the next year he grows them with a second branch; the third year both branches become doubled and another appears, and so each year as he grows them afresh they are more and more complicated, till at last the whole branched tree grows up in a few months. Now in thus increasing his spikes year by year, he is in his own person most curiously retracing the steps of his ancestors in ages past; for, as we have seen, the first deerlike animals had no horns, then as the ages passed on we find that they had single spikes; later on, their descendants grew antlers of two branches, and later still more complicated ones, so that the race put on little by little those magnificent antlers which now the red-deer and others carry, and meanwhile the various species spread all over the world, except into Australia and Africa, south of the desert.

Still, even the stags have times in the year, before their antlers are grown, when they are comparatively defence-less. There remains yet another branch of the "ruminant" family, even better provided with weapons. These are the antelopes, wild cattle, and buffaloes, for with them the horns never fall off. The reason of this is that they grow in quite a different manner from the stags' antlers. Instead of the bone being laid down by the skin, it grows out as a core from the forehead, and the skin over it hardens into horn as it grows, so that the tip of a bull's horns is the oldest part.

Here then we can have no branching as in the stag, but on the other hand a firm and terrible weapon increasing from year to year; and even the king of the beasts, the lion, when he attacks a large buffalo, is often seriously wounded for his pains. We should not wonder then if these animals had

Fig. 71.

A Buffalo cow defending her calf.—(Livingstone.)

conquered the world wherever man had not destroyed them; but strange to say, they have kept chiefly to the old world, for none have travelled to South America, and only the Bisons have overrun North America with their vast herds. All the rest, buffaloes, wild cattle, antelopes, gazelles, goats and sheep, have made their home in Europe, Asia, and Africa, and a fine time they must have had of it when all Europe was one field of undulating plains and dense forests, and the ancestors of our cattle crashed through the tangled bushes, drank by the silent rivers, or grazed on the wild rough herbage. Then, where town and villages now stand, there must have been scenes such as travellers still relate of Central Africa, where amid dense jungle, magnificent forests, and flat marshy grounds,

> "... the elephant browses at peace in his wood,
>   And the river-horse gambols unscared in the flood,
>   And the mighty rhinoceros wallows at will,
>   In the fen where the wild ass is drinking his fill."

There the huge buffaloes come down in troops out of the forest to drink, while the great hippopotamuses leave their watery bed to feed on the rough grass of the swamps. Not far off, a herd of zebras comes galloping by to drink lower down in the river, startling the large antelopes feeding quietly in the soft green pasture above, for they know that this is the hour when the lions are abroad and will fall upon any straggler with tooth and nail, while the distant howling of the hyænas shows that they would not be far behind in seizing upon any weak or wounded animal. But little does the heavy rhinoceros care for all this as he too tramps slowly along on his way to drink, for with his size and defences he runs but little risk of attack. Thus all the country is alive with large milk-givers, and we realise that when they ruled all over the world, as they still do in Africa, they too must have had their time of triumph and greatness like the great fish or the monster reptiles.

But hush! as we watch this scene a heavy thud, thud, strikes upon our ear, like the tramping of heavy troops upon soft ground. It is the "lords of the forest," the large Elephants, which, after feeding all day in the shady jungle, are coming down to drink and bathe. What, then, is the history of these huge antiquated animals that they have not come into our story as yet? The reason is this: as they stand alone now with their huge flapping ears, their column-like legs and feet, and their long grasping trunk, so they have stood apart from the hoofed animals almost as long as we have any knowledge of them. So far as we can judge by their skeleton, especially the shoulder blade, they come nearer to the gnawers, or rodents, than to any of the large vegetable-feeders. Their legs are awkward and their gait clumsy, for the thigh bones are enormously long and thick, and the toes are enclosed in a thick pad with only the nails to mark them; but above all it is the head and mouth which make so strange a figure. Look at the huge forehead, showing a skull of immense size. This skull would be far too heavy to carry if it were not full of hollows, making a large framework to bear the tusks of smooth white ivory, which grow out from the upper jaw to a length of more than six feet on each side,[170] and weigh sometimes from eighty to one hundred pounds. Surely a wonderful size for teeth, and we shall not wonder that they are the only front teeth that the elephant has, and that they go on growing all his life from a permanent pulp,

170   In the African elephant; in the Indian they are smaller, and the female has none.

Fig. 72.

The Indian Elephant.

like the gnawing teeth of the rodents. But if he opens his mouth you will see that, besides these, he has at the back huge flat grinders, one, or never more than two, at a time on each side; but those are monsters, with hard enamelled ridges for grinding his food. During his lifetime of about a hundred years the elephant grows six of these teeth on each side, twenty-four in all, the new ones growing up at the back and pushing forward as the old ones wear away.

And, last of all, look at his wonderful trunk; see how it grows out straight from his face, his cheeks merging into it so that he is all nose; and then consider that this trunk, a double-barrelled tube, ending in a fleshy finger opposite to a thick cushion which acts as a thumb, is the elephant's arm and hand, with which he feels and grasps and tests everything that comes in his way. With it he can pick up a crumb or root up a strong tree, gather a leaf or tear off a branch, draw up a gallon of water to squirt over his body when heated with the sun, or suck up the few drops in a puddle when water is scarce; with it he caresses those he loves, as gently as a mother strokes her child with her hand, or uses it to dash his enemy upon the ground, before he pierces him with his tusks or tramples him under foot.

And yet this formidable and delicate weapon is nothing more than a long fleshy nose and upper lip, provided with millions of interlaced muscles, which draw it in every direction, guided by the delicate nerves. If we did not see it, could we have believed that any creature could have gained so much experience, and learned to do so many wonderful things as elephants do, merely by possessing a movable nose?

Yet so it is, for if the elephant stands far above all other vegetable-feeding animals in intelligence and even reasoning power, we can only attribute it to two causes—the long life he leads, and the delicate implement he carries for testing things around him. The strongest of all animals, he has reigned supreme for ages, even the lion or the tiger often meeting a terrible death from his trunk, his tusks, or his heavy feet, if they venture to attack him; while everywhere, during his hundred years of life, he has handled and tested and tried every object he has come near with his fleshy trunk, till now when we examine his brain we find that though small for so large an animal it is folded and refolded into those curious convolutions which are always found in highly intelligent animals.

For many long ages this education must have been going

on; for already, when the monkeys and opossums were playing about the trees in England, an ancient elephant called
the Mastodon, having four tusks, was roaming over Europe,
Asia, and America; while soon after, the hairy Mammoth,
kept warm by his shaggy coat, wandered right up into the
snows of Siberia and the extreme of North America, and
often met his death in the ice, and true elephants ruled the
world in Europe and India, continuing down to our day. All
these had the same delicate trunk, and gained experience as
they wandered over the wide world, till some have become
extinct and others have shrunk back into the dense forests
of Africa and India, where they often give proofs of a power
of reasoning which surprises us, and make them seem like
old patriarchs of a bygone time, looking thoughtfully upon
a world which has grown new and strange.

* * * * *

And here we must take leave of the Herbivora, and turn
our attention to that large army of flesh-feeders which we
find throughout all past ages harassing and destroying the
vegetable-feeders on all sides, killing their young, falling
upon the stragglers, the weak and the aged, and keeping
down their numbers by constant persecution. For, since the
whole world is teeming with life, and countless new beings
are coming into existence day after day, there is no creature
on the earth which has not some other creature to prey upon
it. Thus, for example, the whole host of small animals, rats
and rabbits, moles, shrews, and small birds of all kinds,
have their special pursuers in long wiry-bodied civets and
ichneumons, weasels, pole-cats, ferrets, pine-martens, and
paradoxures, which can work their way into a hole, give
chase through the long grass, or climb the trees and feed
on birds' eggs or young birds. There is a vast multitude of

these smaller flesh-eating animals, with teeth so sharp that a weasel will kill its prey in a second by piercing the skull by its bite; and they make sad havoc all over the world among young and weak creatures, while a great many of them, such as the weasel tribe, the pole-cat, and the skunk, are themselves protected from larger animals of prey by their disagreeable smell.

Fig. 73.

The Weasel[171]—a small, long, narrow-bodied carnivorous animal.

Then the birds again have their numbers greatly thinned by the wild cats, tiger-cats, and racoons; while the fox, the badger, and the glutton, do their share in devouring partridges and all ground birds, hares, rabbits, and even lambs and other young creatures.

171   Mustelus.

Fig. 74.

The Egyptian Ichneumon,[172] a long-bodied
carnivore, sucking crocodile's eggs.

Lastly the fish, too, have their pursuers, for the mink and
the otter, though true land animals, seek their food in the
water, the sea-otter giving us a hint as to how such flesh-
eating animals as seals, which are the great fish-devourers,
took to a watery life. But though these smaller flesh-eaters
are spread in great numbers over the world, the civets and
ichneumons only in the Eastern Hemisphere, the racoons only
in America, and the weasels and their relations everywhere,
yet the war they carry on is but little seen compared with

172   Herpestes.

the ravages of their more imposing relations the wolves, the bears, and the lions, tigers, and their kin. For these animals seek their prey among the buffaloes, antelopes, horses, sheep, and hogs, and where they go they leave the track of blood behind them, and appear indeed as ruthless destroyers.

And yet it would not be fair to speak of these larger flesh-feeding animals as if they had worked nothing but evil to their more peaceful neighbours; for how would Life educate her children if she put no difficulties in their way to be conquered, no sufferings to be endured? We saw that in the beginning the vegetable-feeders were neither so strong, so intelligent, nor so swift of foot as they are now, while the flesh-feeders were not nearly so well armed for destruction as the tigers and lions of to-day.

It was in the long long struggle for life that the animals with the largest and strongest horns got the upper hand, that the swiftest horses or antelopes survived and left young ones, that the best climbers baffled their hungry pursuers, while the most intelligent and cautious feeders learned to herd together and watch for danger; while we must remember that it is more often the sickly, worn-out, and diseased animals that fall a prey to the devourers, and their life is ended far less painfully than if they dragged themselves into some hole to die. And so, too, on the other hand, with the flesh-feeders themselves. It was no wanton cruelty that taught them to hunt for prey, to creep stealthily along and leap upon their victims, and to take advantage of the weak and feeble. It was pressing hunger and the necessity of providing their young ones with food; and they, too, have often suffered in the struggle; so that it was only the strongest, healthiest, and best armed, that won the victory and were able to bring up their children.

Fig. 75.

The Wolf,[173] showing the dog-like form, and long mouth full of teeth.

Moreover, it is quite a mistake to suppose that the greater part of the life of a lion or a wolf is spent in killing and destroying, any more than ours is because we eat beef and mutton. The Lion, at any rate, never attacks an animal unless he is hungry, and even the wolf, generally considered so cruel and bloodthirsty and pitiless, spends the greater part of the year in some quiet place in the mountains with wife and cubs, only hunting for their daily food (though sometimes he is guilty of killing more than he needs), and playing, gambolling, and resting the remainder of the time.

It is when winter comes, and the young ones are stronger and food is scarce, that he grows wild with hunger, and starts off, with a number of others, to scour the forests, so that the animals fly in terror as they hear the howling from afar; and even the traveller, driving his sledge across the

173   Canis lupus.

snow, urges his frightened horses to their utmost speed, since, with a pack of hungry wolves, even if he has firearms, his life is at stake.

The Wolf, with his relations, the foxes and jackals,[174] is the form of flesh-eating animals which has become least altered from the general type of milk-givers. He has the slim form peculiar to flesh-eaters, but the claws of his feet cannot be drawn in like those of tigers, nor has he those powerful hindquarters which enable them to bound and leap, or the strong paw and fore leg with which they give the death-blow to their prey. Moreover, his face is long like a sheep's, and his jaws are full of teeth, some of which are blunter than the tiger's teeth, and more fitted for grinding, for wolves and dogs are omnivorous. But then, on the other hand, he is not so much of a vegetarian as the bears, nor has he their clumsy gait and cumbersome body, for he walks upon his toes and not his flat foot; lastly, his front teeth are large and sharp, and his fangs strong, for they are his chief weapons, and he uses them with wonderful effect. He is essentially a running animal, and chases his prey, rarely leaping on it but tearing it down with his teeth. Strong as he is, he seldom attacks an animal larger than himself, except when he has companions to help him, and then, indeed, he makes little account of a horse or a buffalo, for combination and co-operation are the great strength of the wolf tribe. Even their cowardly cousins the Jackals hunt in packs when they attack living animals, feeding at other times on offal and the remains of the lion's feast. Yet such is the power of numbers that there is no part of the world, except a few

---

174   These are united in one family, the *Canidæ* or Dog family; but this name is unfortunate, as there are no original wild dogs, only those which have run wild from man. Dogs are now almost certainly shown to be descended from wolves and jackals.

islands, where some member of the wolf family is not to be found. In Northern Europe, Asia, and North America, the common wolves and the prairie wolves hunt in large packs, and in South America the Red Wolf takes their place. In Africa and India the jackals wander with their dismal howl; and even in Australia the wild Dingo dog, probably brought there long ago by savage man, is the terror of all peaceful creatures.

Nor must we forget the cunning clever Fox, with his keen face and bushy tail; for he, curiously enough, is the only one of the wolf family which always hunts alone. The reason of this probably is that he contents himself with small prey—birds, rabbits, and game; while his burrowing habits, his cunning, and his night-hunting, enable him to escape destruction. He is one of the most subtle and knowing of animals except, perhaps, the jackal; and the fact that the pupil of his eye expands and contracts like a cat's, especially fits him for night-work. So, although he has only himself to depend upon, his race has spread from the Arctic regions, where the Blue Fox wanders over the frozen sea to eat dead seals, down to Africa where the tiny Fennecs feed upon dates, and South America where the Gray Foxes follow the jaguar, as the jackals in Africa do the lion.

And now, does it not seem strange that from a family so fierce and bloodthirsty as the wolf family, our own true, faithful, large-hearted dog should have sprung? But do not let us be too hasty. Remember that this hunting and killing is not for pleasure but for daily bread, and that the wolf and jackal at home are good, tender, and loving parents; and, moreover, that they have both of them been tamed, and shown great affection to man.

Surely we wrong the animals when we call bad men

"brutes," for men love and forget, but a dog will die on his master's grave, and a tame wolf, whose mistress went away, pined and grieved till she returned, when, on hearing her footstep, he bounded to meet her, and springing up upon her, fell back dead,—his faithful heart had burst with the shock of joy.

And then, also, we must remember that the family of the wolf is the only one among the carnivora in which the animals hunt in packs, so as to learn sociable habits and to obey the will of others. And here, perhaps, we have the reason why, though we have tamed the cat and brought her to our homes, she still remains half-defiant, and can never be taught to work for man; while the dog, on the contrary,

Fig. 76.

The Tiger.

Showing slim body, muscular thighs, strong front legs and paws, and short face with large teeth, all with sharp edges, especially one (*the carnassial*), near the back in both jaws.

has become our obedient servant, and will tend our sheep, guard our homes, and defend our lives.

Loving, and affectionate indeed, as she is, yet the cat will probably never entirely lose the free untamable spirit of her tribe, for if we search the whole world over we shall not find a creature better fitted for a hunter of prey than the wild cat, the lion, or the tiger. Gentle and loving at home with the wife and little ones, patting with soft paws in which the claws are hidden, and doing no harm to any one till food is needed, yet when they are once out on the chase we see that every part of their structure is of use in approaching and overcoming their victims.

Look at the Tiger as he moves along, crouching to spring upon his prey. Here we have no round barrel-shaped body, with a tight-fitting skin, as in the horse and ox, but a slim slender-waisted animal, which is lithe and nimble, because feeding on nourishing flesh he can do with a small stomach and short digesting tube. So, too, his loose hanging skin, forming a flap under his body, saves him from wounds in his adventurous life, for, when seized by teeth or claw, this skin wrinkles up, so that even if a good grip be taken the tender flesh underneath may escape. This flesh itself is firm and solid, being made of powerful muscles, while the cords or tendons of the body are so thick and strong that he can kill an ox with a blow of his paw; and under this flesh again are bones polished like ivory, far more compact and firm than those of most animals, and bound together by strong ligaments, the rounded joints moving smoothly upon each other and causing those graceful movements which enable him to creep stealthily and spring upon his prey. Lastly, the tips of his toes, upon which he walks, are clothed underneath with a soft pad which breaks his fall when he leaps, and makes his footfall silent as he creeps through the jungle; while, nevertheless, he has sharp claws hidden within to strike when needful.

These movable claws are indeed peculiar to the cat or feline tribe (though the civets and ichneumons can draw theirs in half), and they are caused by the second joint of the toe being grooved, while the end joint, curved and covered with a horny claw, is drawn back by a strong elastic band (*l*) till it lies in this groove so that the outgrowing skin of the toe covers it. There it remains so long as it is not wanted; but when the animal bends its paw to strike, another band or tendon (*t*) *under* the toe is tightened and the claws are thrown forward, burying themselves in the flesh of the victim.

So in shape, in limbs, and in claws, the tiger, the lion, and their relations, are the perfection of hunting animals; and when we examine the well-formed head set upon the strong neck, so that it can turn widely from side to side, ever on the watch, we see that here too everything is fitted for the work. Not only are his ears so quick of hearing that the smallest rustle in the grass startles him at once, while his large round eyes have a special reflecting mirror at the back to catch the faint rays of evening light when he prowls abroad, but the whisker-like tufts on his face are so provided with nerves at their base that when he raises them they are the most

Fig. 77.

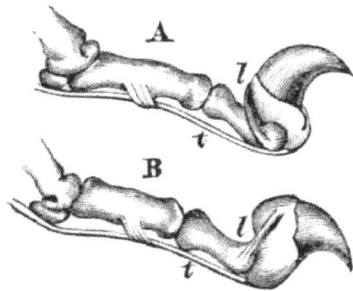

Claws of the Cat or Tiger.
A, claw held back by the strong ligament *l*; B, claw pulled forward by the tendon *t* being drawn back, so that *l* is stretched out.

delicate feelers to guide him in the dark. Then, instead of
the long narrow face, flat teeth, and sideways-moving under
jaw of the horse or ox, we find that he has a large broad
brain-case with a well-formed brain within, and a short face
with rough bony ridges upon it, to support powerful muscles
which move the lower jaw *up and down*, so as to mince the
food, and even crush solid bones.

Such a small mouth cannot hold many teeth, and the front
ones, though sharp and pointed, are small, for the tiger does
not fight with his teeth like the wolf, but strikes with his
heavy paw. But the eye-teeth are immensely large, strong,
and dagger-like, to hold the prey and tear the flesh apart,
and all the double teeth behind, especially the last bottom
tooth and the one to match it above, have very sharp cutting edges, so that, when the two jaws work against each
other they divide the flesh like a pair of shears. Lastly, his
tongue is not soft and fleshy, so as to serve for tasting, but
very rough, and covered with horny pimples which serve to
rasp the flesh from the bones of his prey.

Thus, in all the animals of the cat tribe, such as the lion,
the tiger, the jaguar, and their relations, every part of the
body has become fitted to help them in the work of destruction; and even their near relation the Hyæna, though he
cannot keep his claws sharp by drawing them in, nor leap
so well because his hind legs are short, makes up for this by
his immensely strong jaw and conical teeth with which he
attacks his prey, instead of using his paw, and which serve
him to split open even the strongest thigh bone of a horse
or ox, or to gnaw the ends to extract the marrow.

With all these advantages, we shall not wonder that the
feline family and their near relations were the rulers of the
forests and plains and mountains till man came to conquer

them, or that lions and large cats, something like those living now, together with the fierce sabre-toothed tiger (Machairodus), roamed over Europe, Asia, and North and South America, where the crowds of vegetable-feeders offered them plenty of food. They were even numerous in England, where the lion chased the elk and the wild cattle, before he was driven back to Africa, Persia, and Bengal. No doubt in those days he scraped out his den in the valley of the Thames, as he still does in some quiet spot in the African plains where he hunts alone, except when his little ones are born, and then for some time he lives with his lioness, helping her to provide for them, and taking out the cubs as soon as they are a year old to teach them to hunt, to leap upon their prey, and to strike it with their paw, educating them like a true father in getting their living. And when they are three years old, the young lions will go off and meet together, two or three in a party, till in the spring each one seeks a wife for himself, having many a fierce battle with other lions before he can win her, and finding then the use of his thick mane in protecting his neck from the teeth of his rivals.

So the "king of the beasts" lives

> "On the mountains bred,
> Glorious in strength;"

for though by no means so large as people generally imagine, compared to the buffaloes, or horses, or large antelopes which he attacks, yet his immense strength generally secures him the victory over all but the rhinoceros and the elephant, and he feeds in a royal manner, sharing his hunting grounds only with the leopard, and leaving the remains of his feast for the hyænas and jackals following in his track.

Then just where his reign ends in Bengal, that of the

tiger begins, that splendid and ferocious cat, larger even than the lion, which spares no animal, and will fight till death even with those stronger than himself. When we see our own house-cat playing with a mouse, striking at it, letting it escape, and at last giving it the final grip, we are watching in miniature the cruel game which is played in the dense jungles of Asia by the tiger with the antelopes, young buffaloes, and other terrified animals. Yet when we see the mother cat caressing her little ones, this too is true to tiger life, for though the father does not watch and care for his children as the lion does, the tigress loves them with the utmost devotion, and attacks all who come near them, dying sooner than forsake her cubs.

So in Africa and Asia the lion and the leopard reign, while the tiger is confined to Asia, ranging up to the snowy regions in the Caucasus Mountains and Mantchuria, where he is covered with a warm coat of hair. Yet all these animals have but a small kingdom now compared to olden times; and man has so cleared the ground in other parts of the world that we must travel away to South America to find the other large felines, the fierce Jaguar and Puma. There the jaguar, second only in strength to the tiger, carries all before him, making havoc among the peccaries and the herds of wild horses, and even fishing in the rivers for turtles and fish; scooping the turtles out of their shell with his sharp claws, and conquering every animal except the great ant-bear in whose embrace he has been found dead after he had also killed his enemy. The puma, meanwhile, contents himself usually with smaller prey,—sheep and rheas, opossums and monkeys, for he can climb like a cat, and passes much of his life in the trees. Thus, though the cat family wander over the whole earth, the larger kinds live chiefly in the warm parts of the world where life is luxuriant and man has not yet driven them out.

But these are not all the wild flesh-feeders. There remains a third group—a lazy, easy-going, lumbering group, which, though they spread from the equator to the poles, have taken chiefly to temperate and colder regions for their home, to mixed food for their nourishment, and have gone off on a line as far from the wolves on the one side as the lions have on the other.

Fig. 78.

Polar Bear[175] and Walrus.[176]
Showing how the Bear walks with the heel flat on the ground, and the Walrus also.

175   Thalassarctos (ursus) maritimus.
176   Trichechus rosmarus.

This group is the Bears, and it is a very curious one in many ways. For, in the first place, though they are large and strong animals, they have very much given up eating flesh-food, and have taken to berries and acorns, fruits, vegetables, and honey. To get this last they even climb the trees to dig out the comb with their paw, trusting to their thick shaggy hair to protect them from the stings, which, however, they sometimes receive rather heavily on the nose.

A glance at a bear's mouth will tell at once that he is partly a vegetarian, for his hind teeth are smoothed down, and as he eats he can move his lower jaw slightly from side to side, so as to chew vegetable food. Even the Polar Bear, which eats little else but fish and seals, has these same grinding teeth, and he can be fed for a long time upon bread; while it is found that he keeps in better health when in zoological gardens if he has some grass occasionally. Still it is only the Sun Bears and Sloth Bears in India and Malacca which never eat flesh, for the Bruin of our northern countries often varies his food with deer or sheep, and grows more ferocious and flesh-feeding as he grows in years. It would almost seem as if his very laziness and awkward gait may have led him to take to vegetarianism as a convenient change, when animal food was not handy. For though a bear can trot along at a good pace, yet his heavy lumbering body and long foot with the whole heel touching the ground[177] (see Fig. 78), make him decidedly not well fitted for a hunting animal.

How different he looks from the slim wolf running on the tips of his toes, and the graceful tiger bending his long hind legs for a leap! Yet he is a formidable animal too, for his muscles are tremendously strong, and his firmly-planted foot enables him to rise upon his hind legs and give that

177   Plantigrade.

deadly embrace which drives the breath out of the body of his victim.

The wolf attacks with his teeth, the lion strikes with his paw, but the bear hugs his enemy to death; and here his long stiff claws serve him well, for though he cannot draw them in to keep them sharp, yet they are rough and jagged, and inflict dreadful wounds. The great Grizzly Bear of America, which is sometimes nine feet long, and strong enough to drag along the carcase of a bison, sticks his front claws into his prey while he tears the flesh with the hind feet; he is the only one, except the polar bear, which lives principally upon animal food.

In fact, the bears take much the same place in the animal world that heavy phlegmatic men do among ourselves; easy-going, but dangerous if roused, they seem to have succeeded in life more by accommodating themselves to things as they have found them, than by conquering and taking by force like the wolves and tigers. Thus a bear roams leisurely through the thick forest, for few animals care to meddle with him and he feeds wherever food comes easy, especially in the autumn when fruits abound and he can grow fat; and then he lies down to sleep in a cave or hollow tree, or in a nest of moss and leaves, till spring comes round again. Why should he trouble himself to struggle with difficulties? Unless, indeed, food is scarce, and then he sometimes has an uneasy winter, or attacks animals he would otherwise leave alone.

But if once he is roused, or if a she-bear is afraid that her cubs may be attacked, then you see that under the lazy good-nature there is plenty of pluck and ferocity. He would rather be let alone, for he looks upon life as a thing to enjoy and take leisurely, but if you will have a struggle then he will see who is master. And this kind of philosophy, somewhat easy for strong powerful creatures, has stood Bruin in good stead, for he has spread over all countries where there are thick forests, except Africa and Australia; and with his great strength and shaggy coat must have been very safe from attack till man came to annoy and worry him.

Even the polar bear, living amidst perpetual snow and ice on the shores of Spitzbergen, Nova Zembla, and Green-land, has not, on the whole, a bad life of it, for he is master of the situation, and conquers and devours even the tusked walrus. The polar bear is a most interesting animal, because he shows us the bear tribe becoming adapted to a watery life. His body is much longer and more flexible than that of most bears, giving him the power to twist and turn in the water, as he swims with strong broad feet; and his long neck, narrow head, and small ears, are all fitted for a watery fish-ing life, while he fights entirely with his teeth and does not hug his prey. Again, the soles of his feet, instead of being bare, are covered with long stout hairs, giving him foothold upon the slippery ice, over which he travels very quickly, climbing up from time to time on the icy hummocks to see where seals are to be found, or to scent a dead whale from afar. He is an inveterate seal-hunter, chasing them in the water or out of it with equal ease and great cunning, though they are quick too, and often escape him just when he thinks he has caught them. It is when they are asleep with their noses upon the ice or the land, that he has his best chance, for then he will swim warily behind them, coming up close, till, even if they wake, they have no choice but to be killed where they are, or to leap out on the solid ice where he will soon overtake them.

The polar bear, unlike his brown cousins, fishes and hunts all the winter through, and it is only the mothers which take refuge in caves hollowed out of the snow, where their little ones are born in early spring, and nestle down by her side in their icy home. And when the cubs can run, both father and mother care for them with true devotion, defending them against all attacks, and pushing them before them when pursued, even going so far as to take them in their teeth and swim away with them when they cannot otherwise save them.

So we see that the polar bear has become more than half a water-animal, and gives us the first hint that some milk-givers may take to a thoroughly sea life. Neither among the wolves nor the felines do we find any animals taking entirely to the water; but in the weasel family, which comes near to the bears, we have the otters, and among the bears themselves their polar cousin, which reminds us that there is another great division of flesh-feeders which we must study in the next chapter—the walruses, seals, and sea-bears, the porpoises, dolphins, and whales, which with finned paddles have struck out quite a new line of life, and imitated the fish so well that they are often wrongly classed among them.

EVROPE IN THE AGE OF ICE

CHAPTER XI.

HOW THE BACKBONED ANIMALS HAVE RETURNED TO THE WATER, AND LARGE MILK-GIVERS IMITATE THE FISH.

"ON revient toujours à ses premiers amours," says the French song. But who would have thought that, after rising step by step above the fish, and tracing the history of the backboned animals through their development in the air and over the land, till we brought them to a stage of intelligence second only to man, we should have to follow them back again to the water and find the highly gifted milk-givers taking on the form and appearance of fishes? Nevertheless it is so, for seals and whales are as truly flesh-eating milk-givers as bears and wolves; nor are they much behind them in intelligence, for we all know how teachable and affectionate seals and sea-lions are, while what little is known of the life of whales shows that they are devoted mothers, and their well convoluted though small brains are a proof that they are by no means wanting in intelligence.

Yet the whales and dolphins, at any rate, have not only adopted a sea life, but have limbs so like a fish's fins that we can scarcely call them by any other name, and they are so completely water animals that they cannot even return to the land.

Now we should be quite puzzled to account for such curious forms as these warm-blooded animals, half transformed into fish, if it were not that we know of several land animals belonging to different groups which have gone part of the way towards a fish life. Thus among the reptiles we have the oceanic turtles and the sea snakes; among birds the penguins, whose wings have almost become fins. Then among the milk-givers we have the web-footed Duck-billed Platypus, the Yapock or web-footed opossum of South America, the Desman and the Beaver, the Polar Bear, and last but not least the Otters, web-footed animals nearly allied to the weasels, which seek their food entirely in the water.

The common Otter of Europe and America though he moves quickly and actively on land, has webbed toes with only short claws standing out beyond the swimming foot, and he spends the greater part of his life in the river, making his home in a hollow of the bank beneath the overhanging roots of trees. There he may still be seen in many of our English rivers, his soft brown fur shining as he swims along, diving under water for a fish, which he brings out on to the bank to eat, holding it in his fore paws.

But there is an otter which has deserted the old land life much more completely than this, for the great Sea-Otters of the North Pacific, about four or five feet long (see Fig. 79), never care even to come on shore, but, when they have dived for their prey, turn on their backs and float while they eat it, holding the sea-urchins, crabs, or fish, in their front paws. They even nurse their young ones in the same fashion, dandling them in their arms as they lie face upwards on the sea; and they rear them entirely on the thick beds of kelp off the coasts of the North Pacific Ocean, never bringing them on land.

These sea-otters may be seen in hundreds off the coasts of Alaska and California, basking on the wet rocks, playing, leaping, and plunging in the water, till some alarm makes each mother seize her little one in her teeth and dive under in an instant.

They are twice the size of the River Otter, and in many points more like seals, for though their front paws are short and cat-like, their hind feet are flat flippers, with a long outer toe; their face too is broad and short, and their teeth are neither cutting like the weasels nor flattened like the bears, but covered with rounded knobs, well fitted for crushing crab-shells and the bones of the fish on which they feed.

Fig. 79.

Sea-Otter.[178]—(*From Wolf.*)
Showing the front paws, and the hind webbed feet.

We see, then, that it is quite possible for land-animals
to have near relations specially adapted for a sea life. But
the otter is still distinctly a four-footed creature, with free
arms and legs, and we can trace his connection with the
weasel tribe. It is quite different with the three groups of
real fin-footed animals—the Seals and Walruses, the Manatees, and the Whales. Though we can trace their likeness
bone by bone to the land animals, yet they have become so
different as to show that they must have branched off long
long ago; so long indeed that we cannot even guess at the
relations of the whales, while the seals have only a distant
resemblance to the bear family, and the sea-cows or manatees
to the ancestors of the hoofed animals and elephants. Nor

178   Enhydra marina.

shall we wonder to find the whales so much the most fitted for the sea, when we learn that they were already living in the water when we first meet with the great army of milk-givers (see p. 204) just after the Chalk Period, so that they have probably had a much longer spell of watery life than the seals and sea-cows, whose remains we only find later.

Yet even the seals are so much altered from anything we see on land, that few people would believe at first sight that they have the same skeleton as a bear. We need not leave the British shores to study these pretty creatures, for they still come to the coasts of Wales, Cornwall, and Ireland; while in the Hebrides they may be seen lying fast asleep on the rocks at low tide out at sea, one, placed higher than the rest, keeping awake as sentinel to give warning at the least approach of danger.

But if we begin our study with the common seal we shall be much puzzled, for he is very unlike a land animal. His round neckless body tapering away to the tail, where the hind flippers stretch out behind like fish's fins, reminds us far more of a tunny fish than of a four-footed milk-giver; while the front flippers, coming out so finlike from his side, give us very little idea of legs (see Fig. 81). No! in order to compare these fin-footed[179] creatures with land animals we shall do far better to travel up to the Aleutian Islands at the entrance of Behring's Straits, and visit the Fur Seals and Sea Lions, from which we get our seal-skins, and the Walruses which sometimes lie there sleeping on the rocks, though their real home is farther north within the Arctic Circle, round the coasts of Nova Zembla, Spitzbergen, and Greenland.

179   Pinnipedia.

Fig. 80.

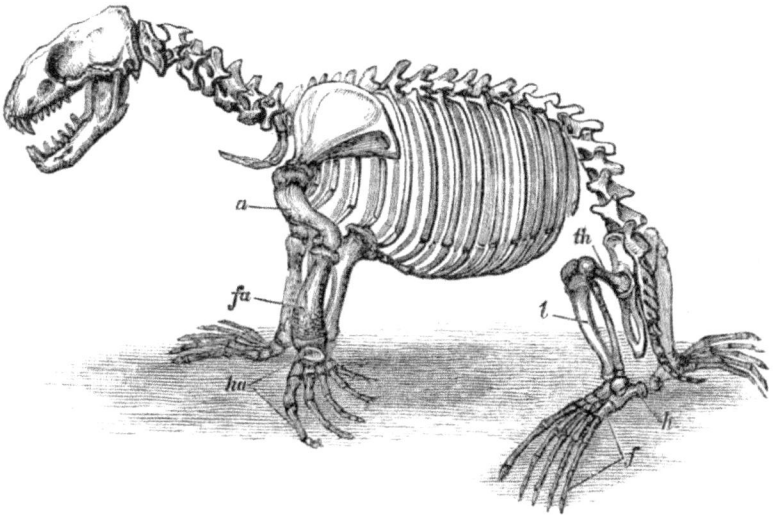

Skeleton of a Sea Lion.
Showing how the whole foot rests on the ground, as in the Bear Family.
th, thigh; *l*, leg; *h*, heel; *f*, foot; *a*, upper-arm; *fa*, fore arm; *ha*, hand.

These creatures, although they have "flippers," and are truly fin-footed, are much more like land animals than the smaller seals, for they plant their whole foot on the ground as a bear does, and walk, or, more properly, "flop along" on all fours. A mere glance at the skeleton of the sea lion, which is one of these higher kind of seals with a slight outer ear,[180] shows that it is a four-footed animal, with five toes to each foot, the great toes and the thumbs being the largest. We can see distinctly the short thighs and the long shanks, which give the hind flippers their lanky appearance, and we see, too, the broad stumpy arms, which give such strength to the front flippers in swimming. For the eared seals and walruses use their fore flippers very much in the water, while

180   Otariidæ (*ous*, *otos*, an ear), eared Seals.

the true seals swim almost entirely with the hind flippers, and use the front ones chiefly for guiding themselves.

And now if we turn to the living fur seal we find that the reasons are twofold which make us forget that his limbs are legs. In the first place, the skin of his body comes down very low over his arms (see Fig. 81), while the hand is encased in skin, with only mere traces of nails upon it Then as regards his hind legs, not only are the feet made into flippers, in which the toes are joined by a loose flexible skin, so that they can move them freely when swimming, but the legs themselves are strapped back by a skin passing right across his tail, so that his thighs are kept flat against his side, and only the lower part of the legs has power to move. We lose sight, then, of the limbs, and see very little more than the feet, which are disguised by being turned into flippers.

Now if we once think what is the object of a seal's life, this curious change in its body is at once explained. For seals are the hunters of the sea; fish-food is to them what flesh-food is to lions, wolves, and bears, only that they have a much wider field to hunt in, for they have the whole ocean for their feeding ground, and no one to dispute it with them but the sea-otter in places near the land, and the porpoises and other fish-feeding whales out at sea. In consequence of this we find seals of some kind in almost all parts of the world, except the Indian Ocean, though they evidently prefer the cooler regions. Even the large sea lions live in the North Pacific, as far up as the Aleutian Isles, and in the South Pacific down to the Falkland Islands and Kerguelen's Land, and play about the shores of the Cape, New Zealand, and Australia.

Fig. 81.

A Fur Seal,[181] one of the Sea Lions; and a common Seal.[182]
Showing how the Sea Lion walks on the flat hind feet, while the seal's flippers lie back in a line with the body; note also the absence of an external
ear in the seal.

They have evidently been very successful in exchanging
flesh-feeding for fish-feeding, and if we consider for a moment
what changes a four-footed land animal would wish to make
in its body in order to swim and dive in the water, we shall
see that these changes have taken place in the seals.

First, a flexible body is required to wind and twist rapidly
in the water, and this the seal arrives at by having the cushions of gristle between its joints very large and thick, while
even its ribs are joined to its back by gristly rods, making its
whole body very lissom. Next, a small head, offering little
resistance to the water is an advantage, and this we find in
all seals, while the short neck and extremely sloping narrow
shoulders well encased in fat, make the body slope away gently with no jutting angles, but a round smooth surface from
head to tail where it narrows like the tail of a fish. The next

181   Callorhinus (Otaria) ursinus.
182   Phoca vitulina.

step is to do away with long angular arms and legs, which would impede it in diving and swimming, and here the seal meets the difficulty, not by losing its leg and arm bones, but by having them so shortened and encased in the skin that only the useful broad flippers are free, while the hind legs are set upon a very narrow hip joint (see Fig. 80), so that they bend backwards and work close to the body. Lastly, such a warm-blooded animal would want clothing to prevent it from being chilled in icy cold water, and here we find two protections. First, under the skin is a layer of oily fat, which, while it reminds us of the fat accumulated by bears before they settle down to their winter's sleep, has become in the seals a dense oily mass, acting like a thick blanket in keeping up the warmth of the body; and secondly, the seal, like its distant relations the bears, has a dense furry covering, and over this a number of coarse long hairs, which give it that shining oily look we notice in all seals. No doubt every one has wondered, when watching seals in zoological gardens, where the fur can be which makes our sealskin muffs and jackets. The fact is, that this under fur is quite out of sight in the living seal, being covered by the coarse hairs; but if we could turn these aside, even in common seals, we should see the soft undergrowth beneath, and in the fur seals it is much thicker. Now the roots of these coarse hairs are deeper in the flesh than the roots of the soft undergrowth, and when the uppermost layer of the skin on which the fur grows is sliced off, the coarse hairs are cut away from their deep roots below, and can then be pulled out, leaving only the fur behind.

The seals then, while they are in all main points constructed like land animals, have gained many advantages, not by having new parts, but by the old ones becoming so modified as to make them admirably fitted for a watery life; and when we add that they have large eyes well adapted for seeing under water, keen ears with little or no outer ear, which would be useless, but a very acute hearing apparatus within, and nostrils which will close firmly and keep the air in and the water out when they dive, we must acknowledge that they make good use of all parts of their body. Indeed, their breathing apparatus is the most curious of all, for they can remain under water sometimes for twenty minutes, and meanwhile the circulation of their blood is probably controlled by large reservoirs in the veins, which prevent it going back to the heart and lungs till it can be purified by fresh breath.

Now, if all these changes from a land to a water-frequenting animal have been made gradually, we shall expect to find some forms less altered than others, and so it is. The Walrus, which is not a seal, but a creature with a thick hide having no fur and only a few scattered hairs upon it, and long tusks in his mouth, is much more of a land-animal than the seals. He passes a great part of his life sauntering along on the low shores of the Arctic seas, digging up mussels, cockles, and clams with his long canine teeth or tusks; and in accordance with this we find that his hind legs are much freer than even those of the sea-lions, for the skin binding them to his body is broader and his hips are stronger, so that, as he throws his front flippers forwards, he can also throw out his feet and walk on all fours in a strange straddling manner. He is remarkably fierce and strong, and Captain Scoresby caught one once in the act of killing and eating a large narwhal, so that they are evidently not afraid of attacking even large animals. The walrus is even said to stand at bay on shore and fight his great destroyer the polar bear, throwing up his head so as to strike forcibly with his sharp tusks, but in this battle he is generally defeated. His tusks alone would suggest that he lives a good deal on land exposed to dangers, for his more aquatic relations the seals are without tusks, and though their teeth are sharp enough, and they fight among themselves, yet their way of escaping the great tyrant of the ice-fields is to slip into the water.

Beyond his tusks, and the fact that by sleeping many weeks on the ice in autumn he reminds us of the bears, the walrus's life is not very interesting. They live in large shoals in the Arctic sea, climbing the rocks and ice with the help of their tusks, which they drive into the crevices and so haul themselves up. During the colder times just before our own, they came down into much lower latitudes than now, and we find their bones as far south as England in Europe and Virginia in America, and even in our day one has been seen off the west coast of Skye; but we know very little of their daily life or how they bring up their young ones.

Of Fur Seals and Sea Lions, however, we know a good deal, and a singular history it is. They spend the greater part of the year in huge shoals in the sea, rising and falling, gambolling and diving in the water, feeding on the fish, and probably migrating from colder to warmer seas in the winter from either pole. But the interesting time of their life is in the spring, when the northern eared seals have often been watched as they come to the shores of the Aleutian Isles to bring up their families.

For then begins the fight which seals shall get the most wives. Early in May the fathers begin to arrive—strong old seals, which have gone through the battle many years before and know the rules. They are huge fellows six or seven feet long, with enormous eye-teeth and cutting teeth next to them, which together grip like a vice. They come up at first singly and then in greater numbers, swimming powerfully and laying hold of the rocks with their flippers so as to haul themselves up on land, taking the best positions they can find on the edge of the water to watch for the arrival of the mothers. Yet still more and more fathers arrive as time goes on, and these are obliged to go farther inland, for all the shore stations are soon occupied, and each sea lion defends his own plot of ground with tooth and flipper.

Fig. 82.

Sea Lions gathered on one of the Pribylov
Islands, watching for wives.

Thus, in about a month's time, from the shore right inland, the whole island is covered with male seals. And now the mothers arrive, coming to the islands that their little ones may be born. They are very much smaller, not much more than four feet long, lighter in colour than the fathers, gentle and inoffensive; and as they swim up to the island each father seal tries, by coaxing, pulling, and tugging, to persuade a mate to come on to his rock. If he succeeds he has then to keep her, for the sea lions behind, which cannot reach the sea, are on the watch to steal her.

Now he might make quite sure of his prize if he would be content with one, but he wants several; and the next young mother swimming up calls off his attention, and while he is courting her his neighbour behind tries to carry his first wife away, lifting her by the back of the neck as a cat does a kitten. Then often a terrible battle begins, and the poor mothers are pulled hither and thither till one male seal secures her, and then the whole thing begins again. This constant fighting and lovemaking go on for several days till all the sea lions have wives—those on the shore many, those behind perhaps very few. Then all settle down quietly, the little sea lions are born, bleating like young lambs, and family life begins. But the peace does not last long, for no sooner are mothers able to leave their little ones than the old contest begins again, and happy the father who can keep his wives together through a whole season!

And now comes the most remarkable point. As a rule, seals are immense eaters, and they become very fat. But from the time that the fathers land upon the rocks till they go back to the water after about two months, they have never been known to leave their position to take food, so busy are they defending their wives. And when the two months are over, during which the little ones have been trying their strength in the waves and learning to swim, the fathers, which have grown thin and meagre, having used up all their fat, swim away and do not come back. The mothers, however, with the children, and those young bachelors, which have not yet taken wives, remain on the islands sporting and enjoying themselves till autumn, when they, too, start off for the open sea till spring comes round again.

Such is the history of the eared seals. And now that we have studied their form, and seen that their skeleton is like that of other animals, though their arms and legs are disguised as flippers, we shall understand our own home seals better; for the chief difference between them and the higher seals is merely that their front legs are much shorter, and that their hind legs are turned back so as to lie in a line with the body (see Fig. 81), while they are closely bound to the tail down right as far as the heel, so that they cannot throw their hind flippers forward nor use them in walking. Thus they have become still more completely aquatic animals, using their hind legs entirely in swimming, when they serve as great oars, working something like the screw of a steamer. The consequence is that they are terribly awkward on land, though they get along very fast by jerking their body forward, or sometimes by dragging themselves by their front flippers.

This, however, matters very little to them, for their home is the sea. True, they may often be seen lying asleep on sandbanks or on rocks jutting out of the water, but they rarely venture far up the land, always remaining where they can slip back into their true home at the least alarm. So they live in the seas almost all over the world. They may be known from the higher seals chiefly by their want of outer ears, their backward-turned legs, and their feet with both the great and little toes larger than the inner ones; but their life is much the same. Some live near our own shores, especially in Scotland; some are peculiar to Australia and New Zealand; others crowd the icy seas of Greenland, sleeping

in large herds on the ice-fields, where the polar bear makes
them his prey; while others again live on the pack ice round
the South Pole, the huge Elephant seal, with its long tapir-
like nose, basking on the shores of Kerguelen's Land and
the islands of the southern seas—a monster twelve feet or
more long, with his smaller wives beside him.

* * * * *

Thus the seals are bold ocean lovers, feeding entirely on
animal food, and finding plenty of it in the wide sea as they
roam. But there is another family of warm-blooded animals,
pure vegetable-feeders, which also must have found their
way in distant ages into the water; for they too are milk-
givers, and though they have lost their hind legs, have still
the front legs with all their proper bones, with the hands
turned into flippers.

These animals are the curious sea-cows or Manatees,
which wander under water along the east coast of Africa
and west coast of South America, feeding in the bays and
often up the rivers, on the seaweeds and water-plants of all
kinds; while another kind with tusks, called the Dugong,
feeds all along the shores of the Indian Ocean and Australia.

It is strange that while every child knows something about
seals, very few people have heard of these gentle grazing
manatees and dugongs, the only large vegetable-feeders
of the sea. Yet they are curious, interesting animals, and
seem to be the forms which have given rise to the popular
stories of mermaids,[183] for they suckle their young ones at
the breast, clasping them with their flippers, and when they
raise their heads in the water have something the appearance
of an uncouth mother nursing her child.

---

183   Hence their name *Sirenia*, a curious name for voiceless animals.

Fig. 83.

The Manatee or Sea Cow grazing.

But very uncouth indeed! for they are long barrel-shaped creatures, with a thick skin like the elephant's, with short stiff hairs upon it. Their head is small, with no outer ears, and very insignificant eyes surrounded with wrinkles; their lips are thick, heavy, and covered with short bristles, and above them two narrow nostrils open and close according as they are above or under water. Their front flippers, which are all they have, are long and broad, with faintly-marked flat nails upon them, and behind these their body tapers away gradually into a thin, wide, shovel-shaped tail, not set edgewise as in a fish, but *across* the body, so as to lie like a broad leaf in the water.

Who would think that a creature like this had anything in common with land animals? Yet so it is, for not only do we know that his ancestors had traces of hind legs, but his front limbs are quite as true arms and hands as those of any of the seals. Moreover, he has large broad grinding back teeth like the elephant, and in front he has small cutting teeth as a baby, though these are covered up by the gum as he grows older. In the Australian dugong, however, these teeth continue to grow and form good-sized tusks in the fathers. What, then, is this curious animal? Simply a vegetable-feeder which has become fitted for a watery life—a gentle, peaceable animal, which keeps near the shore and grasps the seaweed with the sides of its upper lip, and then nips it off by a set of horny plates, which grow down from the roof of its mouth, and answer to the rough wrinkles on a cow's palate. They may often be seen together, father, mother, and child, wandering up the river Congo in Africa, or the Amazons in South America, feeding entirely under water, and only raising their heads from time to time with a snort to take in fresh air. In olden times they probably thronged all the coasts on the sea-margin, for a hundred and fifty years ago there was another group of them, the Rhytinas, right up in the cold seas of Behring's Straits, where the vast submarine forests of seaweed afforded them plenty of food. But the sailors found them such good eating, and the fatty blubber on their bodies was so valuable, that they were all killed twenty-five years after Behring first discovered them, and unless some care is taken, the more southern sea-cows may some day be exterminated in the same way.

* * * * *

And now that we have firmly grasped the fact that the seals and manatees, however altered in shape, belong to the four-footed and milk-giving group, perhaps we shall be prepared to understand how it is that the whales[184] are not fish, though this popular delusion is one of the most difficult to overcome. "Do you really mean then," exclaim nearly all people who are not naturalists, "that a whale is not a huge fish?" Certainly I do! A whale is no more a fish than

184 Cetacea—*cete*, a whale.

crocodiles, penguins, or seals, are fish although they too live chiefly in the water.

A whale is a warm-blooded, air-breathing, milk-giving animal. Its fins are hands with finger-bones, having a large number of joints (see Fig. 84); its tail is a piece of cartilage or gristle, and not a fish's fin with bones and rays; it has teeth in its gums even if it never cuts them; and it gives suck to its little one just as much as a cow does to her calf (see Fig. 85). Nay! the whale-bone whales have even the traces of hind legs entirely buried under the skin (see Fig. 84), and in the Greenland whale the hip-joint and knee-joint can be distinguished with some of their muscles, though the bones are quite hidden and useless.

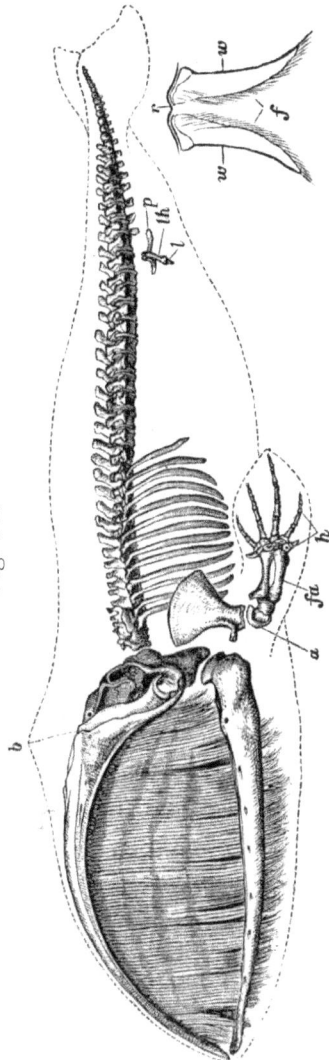

Fig. 84.

Skeleton of a Whalebone Whale (Mysticete), and Section of the Mouth with Whalebone.

b, blowhole; a, upper arm; fa, fore arm; h, hand; p, th, l, small remains of pelvis or hip-bone, thigh, and leg; r, roof of the palate; w, w, plates of whalebone; f, whalebone fringe.

We see then that the whale undoubtedly belongs to the same type as the four-footed land animals, although it branched off into the water so long ago that it may have come from some *very* early milk-giver. But why then has it become so like a fish? For the same reason that the penguin's wings have become so fin-like, and the seal's arms and legs have become flippers, namely, that during the long time in which the whales have taken to a watery life, those which could swim best and float best in the water have been the most successful in the struggle for existence; and as a fish's shape is by far the best for this purpose the warm-blooded milk-giver has gradually imitated it, though belonging to quite a different order of animals.

Fig. 85.

The Humpback Whale[185] suckling her young (*AFTER SCAMMON*).

We saw this imitation already beginning in the seals,

185   Megaptera.

with their bodies sloping off towards the tail and their legs fastened back in a line with the body; but they have not gone so far in this direction as the whales have, since they still have hind legs and furry bodies. The sea cows, on their line, have gone a little farther, for they have lost their hind legs, and their skin is smooth, with very few hairs upon it. But it remained for the whales to take up the best fish-form, the old spindle-shape, thinning before and behind, with the strong fleshy tail ending in two tail lobes, which act like a screw in driving the body along.

Any good drawing of a whale shows at once how admirably these animals are fitted for gliding through the water (see Fig. 85). True, many of them have enormous heads, but these always have long face-bones ending in a rounded point, and even the huge head of the sperm whale (see Fig. 87), eighteen feet long, six feet high, and six feet wide, is rounded off above, and gradually thins away below, like the cutwater of a ship. The eyes are very tiny and so little exposed, that it is difficult to find them; there are no outer ears, though the bones within are large and probably very useful for hearing in water; the bones of the neck are seven, as is the rule among milk-givers, but they are so flattened and firmly soldered together, and so covered with blubber, that there is not even a hollow between the head and the body; while to crown all, the skin is perfectly smooth so as to offer no resistance to the water. Here, however, would be a disadvantage in the loss of the furry covering, since most of the whales travel into cold seas, were it not compensated by the great mass of oily fat or blubber which fills the cells in the under part of the skin, and keeps the whole body warm; and thus the whale, by a covering of fat often as much as a foot and a half thick, solves the problem of a warm-blooded animal, with a smooth gliding body, living in icy water without having its blood chilled.

In every essential for swimming, then, whales are as well provided as any fish, while their immensely strong backbone, and the long cords or tendons running from the mass of muscle on the body to the tail, give them such tremendous power that a large whale makes nothing of tossing a whole boat's crew into the air and breaking the boat in two. But, though they are so far true water-animals, yet they cannot live entirely below as fish can, for they have no apparatus for water-breathing. The outside of their body takes on the appearance of a fish, but inside they have the true lungs, the four-chambered heart, and all the complicated machinery of a warm-blooded animal. Therefore, though a whale may dive deep and remain below to seek its food, yet before an hour has passed even the largest of them must come floating up to the top again, to blow out the bad air through the nostrils at the top of the head, and fill the capacious lungs with a fresh supply. It is then that, partly because of the water which has run into the blowhole, and partly because the rush of breath throws up spray from the sea, we see those magnificent spouts of water which tell that a whale is below. The older naturalists thought that these spouts were caused by the water which the whale had taken into its mouth; but this is not so, and Scoresby, the great Arctic traveller, states distinctly that if the blowhole of the whale is out of the water only moist vapour rises with the breath, while when it makes a large spout this comes from its blowing under water and so throwing up a jet.

If, however, the whale is a simple air-breather and yet swims under water with its mouth open, how comes it that this water does not run down the windpipe and choke the lungs? This is prevented by a most ingenious contrivance. At the top of our own windpipe there is a small elastic lid which shuts when we swallow, and prevents water and food from running down to the lungs. Now, in the whale the gristle answering to this lid runs up as a long tube past the roof of the mouth into the lower portion of the nose, and is kept there tightly, being surrounded by the muscles of the soft palate. The upper portion of the nose cavity then opens on the forehead by means of one or two "blowholes," as the outside nose holes are called; so that when the blowholes are closed the whale can swim with its mouth open and feed under water, and yet not a drop will enter its lungs.

A large sperm whale will often remain twelve minutes or more at the top of the water, taking in air at the single blowhole in the front of its head, and purifying its blood, and then with a roll and a tumble it will plunge down again, and remain for an hour below, trusting to a large network of blood-vessels lying between the lungs and the ribs to supply purified blood to its body and retain the impure blood till it comes up again to breathe.

But the smaller whales and porpoises, which play about our coasts, have to come up much more often, and even when they are not tumbling and jumping, as they love to do, you may see them rising at regular intervals as they swim along, their black backs appearing like little hillocks in the water, as they "blow" strongly from their single nose-slit, take a quick breath in, and sink again to rise a few paces farther on and repeat the process.

Thus provided both with swimming and breathing apparatus, these purely air-breathing animals wander over the wide ocean and live the lives of fish, making such good use of food which cannot be reached by land animals, or those which must keep near the shore, that we shall not be surprised to find that the whale family is a very large one.

But it is curious that the fierce animals of prey among them should be, not the huge whales but the smaller Dolphins, Porpoises, and Grampuses; and this shows how different water-feeding is to land-feeding, since, because the water is full of myriads of small and soft creatures, the sperm whale feeding on jelly-fish, and the large whalebone whale feeding on soft cuttle-fish and the minutest beings in the sea, are those which attain the largest size.

Most people have at one time or another seen a shoal of porpoises either out at sea or travelling up the mouth of some large river, where

> "Upon the swelling waves the dolphins show
> Their bending backs, then swiftly darting go,
> And in a thousand wreaths their bodies throw;"

and though they are small creatures, only about five feet long, they are very good examples of the whale shape, with their tapering bodies, broad tails, and the back fin, which is found in some whales and not in others. Sometimes they swim quietly, only rising to breathe, and then they work the tail gently from side to side; at others they gambol and frolic, and jump right out of the water, beating the tail up and down, and bending like a salmon when he leaps; and whether they come quietly or wildly, you may generally know they are near by the frightened mackerel and herrings, which spring out of the water to avoid them. For the porpoises have a row of sharp teeth in each jaw, more than a hundred in all, and they bite, kill, and swallow in one gulp, without waiting to divide their food, so that they make sad havoc among the fish.

Fig. 86.

The Porpoise.[186]

They are here to-day and gone to-morrow. A few kinds
wander up into fresh water, such as the Ganges and the
Amazons, but by far the greater number range all over our
northern seas, together with their near relations the dolphins,
and the bottle-nosed whales, and the strange narwhal, with
its two solitary eye teeth, one only of which grows out as
a long tusk. All these roam freely through the vast ocean
home, coming into the still bays to bring up their young ones,
which they nurse and suckle tenderly, afterwards moving
off again in shoals to the open sea. There they will follow
the ships, and sport and play, and probably we shall never
know exactly where their wanderings extend, though it
seems that they prefer the northern hemisphere.

Among all the dolphin family the most voracious and

186   Phocæna communis.

bloodthirsty is the Grampus or Orca,[187] which is commonly
called the "Killer Whale," because it alone feeds on warm-
blooded animals, seizing the seals with its strong, sharp,
conical teeth, devouring even its own relations the porpoises,
and attacking and tearing to pieces the larger whales. No
lion or tiger could be more ruthless in its attacks than this
large-toothed whale, which is sometimes as much as twenty-
five feet long and has broad flippers. In vain even the mother
walruses try to save their young ones by carrying them on
their backs; the cunning Orca swims below her, and coming up with a jerk shakes the young one from its place of
safety and swallows it in a moment. Nor do they merely fight
single-handed, for many voyagers have seen them attack
large whales in a pack like wolves, and in 1858 Mr. Scammon saw three killer whales fall upon a huge Californian
Gray Whale and her young one, though even the baby whale
was three times their size. They bit, they tore, and wounded
them both till they sank, and the conquerors appeared with
huge pieces of flesh in their mouths, as they devoured their
prey. How much they can eat is shown by one orca having
been killed which had the remains of thirteen porpoises and
fourteen seals in its stomach!

How strange now to turn from this ravenous hunter to the
huge Sperm Whale, eighty feet long, with a head one-third
the size of its whole body and more than a ton of spermace-
tic oil in its forehead, and to think that this monster swims
quietly along in the sea, drops its long thin lower jaw, and
with wide-open mouth simply gulps in jelly-fish, small fish,
and other fry, thus without any exertion or fuss slaying its
millions of small and soft creatures quietly, as the orca does
the higher creatures with so much battle and strife!

187   Orca gladiator.

For the sperm whale (Fig. 87) must need a great deal of food to feed its huge body. Though it has forty-two teeth in the lower jaw it never cuts those in the upper one, and seems to depend more on sweeping its prey into its mouth than on attacking it. And this perhaps partly explains the use of that curious case of spermaceti which lies in its huge forehead over the tough fat of its upper jaw. For this oil gives out a powerful scent, which, when the whale is feeding below in the deep water, most probably attracts fish and other small animals, as they are also certainly attracted nearer the surface by the shining white lining of its mouth. This light mass is also, however, useful in giving the head a tendency to rise, so that when the whale wishes to swim quickly it has only to rise to the top, so that the bulk of its head will stand out of the water, the lower and narrow part cutting the waves. In this position he can go at the rate of twelve to twenty miles an hour.

But if the sperm whale is curious, as it carries its oil-laden head through all seas from pole to pole, chiefly in warmer latitudes, how much more so are the whalebone whales, which are monarchs of the colder and arctic seas, where they feed on the swarms of mollusca, crustaceans, and jelly animals which live there. For these large whales, though they have teeth in their gums, never cut them, but in their place they have large sheets of whalebone hanging down from the upper jaw (see Fig. 84), smooth on the outside, fringed with short hairs on the inside, and crowded together so thickly, only about a quarter of an inch apart, that as many as three hundred sheets hang down on each side of the mouth of the great Greenland whale.

Fig. 87.

The Sperm Whale.

It is easy to see the use of these whalebones when we remember that this huge whale feeds entirely by filling its enormous mouth with water, and then closing it and raising its thick tongue at the back so as to drive the water out at the sides, straining it through the fine fringes, which fill up all the spaces between the plates and keep back every little shell-fish and soft animal. But it is less easy to guess where these whalebone plates come from, till we look back at the manatee, and remember those horny ridges which it uses for biting, and which are exaggerations of the rough fleshy ridges at the top of a cow's mouth.

Then we have a clue, for each blade of whalebone grows from a horny white gum, being fed by a fleshy substance below much in the same way as our nails are, so that these blades are, as it were, a series of hardened ridges, which grow out from the soft palate, till they become frayed at the edges, and form that dense fringe which is the whale's strainer, upon which he depends entirely for his food.

Explain it as we will, however, it is a most wonderful apparatus. Imagine a huge upper jaw forming an arch more than nine feet high, so that if the whalebone were cleared away a man could walk about inside, upon the thick tongue which lies in the lower jaw fastened down almost to the tip so that it cannot be put out of the mouth. And then remember that this enormous mouth has to be filled with food sufficient to nourish a body fifty or more feet long. Who would ever guess that this food is made up of creatures so small that countless millions must go to a mouthful? Yet the whole difficulty is solved simply by these triangular fringed plates or mouth-ridges (see section Fig. 84, p. 301), covered with horny matter and frayed into minute threads like the horny barbs of a feather.

Nor are we yet at the end of the wonderful adaptation, for while the jaw is only from nine to twelve feet high, the long outside edge of the plates is often eighteen feet long, and for this reason, that if they were only as long as the jaw is deep, then when the whale went fishing with his mouth open the animals would escape below the fringe, while as they now are, he may gape as wide as he will, the long curtain will still guard the passage of the mouth and entangle the prey in its meshes. But what, then, is to become of this great length of whalebone when the animal shuts his mouth? Here comes in the use of the beautiful elasticity of the plates, for the great Arctic whaler, Captain Gray, has shown that as the mouth shuts the lower ends of the longer plates bend back towards the throat and fall into the hollow formed by the short blades behind them, so that the whole lies compactly fitted in, ready to spring open again, and fill the gap whenever the jaws are distended.

With this magnificent fishing-net the whalebone whales go a-fishing in all the salt waters of the world. They are not all of enormous size,—many of them are not more than twenty feet long,—nor have they all such a perfect mouthful of whalebone as the great Polar Whale; but when the whalebone is shorter, as in the Rorqual, and other whales with back fins, the stiff walls of the lower lip close in the sides of the mouth and prevent the escape of the prey; and many of these whales have a curious arrangement of skin folds under the lower jaw, which stretch out and enable them to take in enormous mouthfuls of water, so as to secure more food.

New Zealand, California, Japan, the Cape, the Bay of Biscay, and in fact almost every shore or sea from pole to pole, has some whale called by its name; for these gaping fishers are everywhere, and it is not always easy to say whether the same whale is not called by different names in various parts of the world. In the shallow bays and lagoons they may be found with their newly-born young ones very early in the year; while far out at sea ships meet with them travelling in shoals, or "schools," northwards, as the summer sets in and the Arctic Sea is swarming with life. In fact the Californian gray whales go right up into the ice, poking their noses up through the holes to breathe, and then they travel far away south again into the tropics to bring up their young ones.

And whether large or small, toothed whales or whalebone whales, active as the dolphin and the huge fin-whales or rorquals, which dash through the water although some are nearly a hundred feet long, or lazy and harmless as the

Greenland whale is unless attacked, in one thing all the whale family betray their high place in the animal kingdom. Nowhere, either on land or in the water, can mothers be found more tender, more devoted, or more willing to sacrifice their lives for their children than whale-mothers. Scoresby tells us that the whalers, as means of catching the grown-up whales, will sometimes strike a young one with harpoon and line, sure that the mother will come to its rescue. Then she may be seen coming to the top with it encouraging it to swim away, and she will even take it under her fin, and, in spite of the harpoons of the whalers, will never leave it till life is extinct. Nay, she has been known to carry it off triumphantly, for the lash of her tail is furiously strong when she is maddened by the danger of her child, so that a boat's crew scarcely dare approach her.

And now there remains the question what enemies besides man these strong-swimming milk-givers can have in their ocean home? We have seen that the orca or killer whale will turn cannibal and devour those of its own kind, and the swordfish is said to attack whales with its formidable spear; but these are not their greatest enemies. With many of the whales it is tiny creatures like those on which they feed which hasten their death, for small parasitic crustaceans cover their head and fins, and feed upon their fat, so that whales which have been infested with these animals are often found to be "dry," or to have lost nearly all their oil. And thus we see the tables turned, and while the whale feeds upon minute creatures, it is in its turn destroyed by them.

Nevertheless, as a rule, they probably live long lives, till their teeth are worn, or their whalebone frayed and broken, and their blubber wasted away; and then, it may be after eighty or one hundred years of life, they die a natural death.

Therefore they probably share with the elephant the longest term of life of any of the warm-blooded animals; and though their existence cannot certainly be said to be an exciting one, yet, when undisturbed by man, it is at least peaceful, sociable, and full of family love.

It may perhaps seem strange that we should have taken these ocean-dwellers last in our glimpses of animal life; but in the first place, how was it possible to show how they are truly related to the land mammalia until we understood the structure of these last? And in the second place, we have as our object to see how the backboned family have won for themselves places in the world, and surely there are none which have done this more successfully or in a more strange and unexpected way than the whales, which, while retaining all the qualities of warm-blooded animals, have won themselves a home in the ocean by imitating the form and habits of fish, and so adapting themselves to find food in the great oceans, where their land relations were powerless to avail themselves of it.

WHEN THE COLD HAD PASSED AWAY

CHAPTER XII.

A BIRD'S-EYE VIEW OF THE RISE AND PROGRESS OF BACKBONED LIFE.

WE have now sketched out, though very roughly, the history of the various branches of the great backboned family, and we have found that, as happens in all families, they have each had their successes and their downfalls, their times of triumph, and their more sober days, when the remaining descendants have been content to linger on in the byways of life, and take just so much of this world's good as might fall to their share.

We have seen also that, as in all families of long standing, many branches have become extinct altogether; the great enamel-plated fish, the large armour-covered newts, the flying, swimming, and huge erect-walking reptiles, the toothed and long-tailed birds, the gigantic marsupials, the enormous ground-loving sloths, and many others, have lived out their day and disappeared; their place being filled either by smaller descendants of other branches of the group, or by new forms in the great armies of fish, birds, and milk-givers which now have chiefly possession of the earth.

Still, on the whole, the history has been one of a gradual rise from lower to higher forms of life; and if we put aside for a moment all details, and, forgetting the enormous lapse of time required, allow the shifting scene to pass like a panorama before us, we shall have a grand view indeed of the progress of the great backboned family.

First, passing by that long series of geological formations in which no remains of life have been found, or only those of boneless or invertebrate animals, we find ourselves in a sea abounding in stone-lilies and huge crustaceans,[188] having among them the small forms of the earliest fish known to us, those having gristly skeletons. Then as the scene passes on, and forests clothe the land, we behold the descendants of these small fish becoming large and important, wearing heavy enamelled plates or sharp defensive spines; some of them with enormous jaws, two or three feet in length, wandering in the swamps and muddy water, and using their air-

188   Picture heading, Chap. II.

bladder as a lung. But these did not turn their air-breathing discovery to account; they remained in the water, and their descendants are fish down to the present day.

It is in the next scene, when already the age of the huge extinct fishes is beginning to pass away, and tree ferns and coal forest plants are flourishing luxuriantly, that we find the first land animals,[189] which have been growing up side by side with the fish, and gradually learning to undergo a change, marvellous indeed, yet similar to one which goes on under our eyes each year in every country pond. For now, mingling with the fish, we behold an altogether new type of creatures which, beginning life as water-breathers, learn to come out upon the land and live as air-breathers in the swamps of the coal forests.

A marvellous change this is, as we can judge by watching our common tadpole, and seeing how during its youth its whole breathing organs are remade on a totally different principle, its heart is remodelled from an organ of two chambers into one of three, the whole course of its blood is altered, some channels being destroyed and others multiplied and enlarged, a sucking mouth is converted into a gaping bony jaw, and legs with all their bones and joints are produced where none were before, while the fish's tail, its office abandoned, is gradually absorbed and lost.

The only reason why this completely new creation, taking place in one and the same animal, does not fill us with wonder is, that it goes on in the water where generally we do not see it, and because the most wonderful changes are worked out *inside* the tadpole, and are only understood by physiologists. But in truth the real alteration in bodily

189   Picture heading, Chap. IV.

structure is much greater than if a seal could be changed into a monkey.

Now this complete development which the tadpole goes through in one summer is, after all, but a rapid repetition, as it were, of that slow and gradual development which must have taken place in past ages, when water-breathing animals first became adapted to air-breathing. Any one, therefore, who will take the spawn of a frog from a pond, and watch it through all its stages, may rehearse for himself that marvellous chapter in the history of the growth and development of higher life.

And he will gain much by this study, for all nature teaches us that this is the mode in which the Great Power works. Not "in the whirlwind," or by sudden and violent new creations, but by the "still small voice" of gentle and gradual change, ordering so the laws of being that each part shall model and remodel itself as occasion requires. Could we but see the whole, we should surely bend in reverence and awe before a scheme so grand, so immutable, so irresistible in its action, and yet so still, so silent, and so imperceptible, because everywhere and always at work. Even now to those who study nature, broken and partial as their knowledge must be, it is incomprehensible how men can seek and long for marvels of spasmodic power, when there lies before them the greatest proof of a mighty wisdom in an all-embracing and never-wavering scheme, the scope of which is indeed beyond our intelligence, but the partial working of which is daily shown before our very eyes.

But to return to our shifting scene where the dense forests of the Coal Period next come before us. There, while numerous fish, small and great, fill the waters, huge Newts have begun their reign (*Labyrinthodonts*), wandering in the marshy swamps or swimming in the pools, while smaller forms run about among the trees, or, snake-like in form, wriggle among the ferns and mosses; and one and all of these lead the double-breathing or amphibian life.

In the next scene the coal forests are passing away, though still the strange forms of the trees and the gigantic ferns tell us we have not left them quite behind; and now upon the land are true air-breathers,[190] no longer beginning life in the water, but born alive, as the young ones of the black salamander are now (see p. 81). The Reptiles have begun their reign, and they show that, though still cold-blooded animals, they have entered upon a successful line of life, for they increase in size and number till the world is filled with them.

Meanwhile other remarkable forms now appear leading off to two new branches of backboned life. On the one hand, little insect-eating warm-blooded marsupials scamper through the woods, having started we scarcely yet know when or where, except that we learn from their structure that they probably branched off from the amphibians in quite a different line from the reptiles, and certainly gained a footing upon the earth in very early times. On the other hand, birds come upon the scene having teeth in their mouths,[191] long-jointed tails,[192] and many other reptilian characters. We have indeed far more clue to the relationship of the birds than we have of the marsupials, for while we have these reptile-like birds, we have also the bird-like reptiles such as the little Compsognathus, which hopped on two feet, had a long neck, bird-like head and many other bird-like characters, though no wings or feathers.

190   Picture heading, Chap. VIII.
191   Picture heading, Chap. VI.
192   *Ibid.* Chap. VII.

The birds, however, even though reptile-like in their beginning, must soon have branched out on a completely new line. They for the first time among this group of animals,[193] have the perfect four-chambered heart with its quick circulation and warm blood; while not only do they use their fore limbs for flying (for this some reptiles did before them), but they use them in quite a new fashion, putting forth a clothing of feathers of wondrous beauty and construction, and with true wings taking possession of the air, where from this time their history is one of continued success.

And now we have before us all the great groups of the backboned family—fish, amphibia, reptiles, birds and mammalia; but in what strange proportions! As the scenery of the Chalk Period with its fan-palms and pines comes before us, we find that the gristly fish, except the sharks and a few solitary types, are fast dying out, while the bony fish[194] are but just beginning their career. The large amphibians are all gone long ago; they have run their race, enjoyed their life and finished their course, leaving only the small newts and salamanders, and later on the frogs and toads, to keep up the traditions of the race. The land-birds are still in their earliest stage; they have probably scarcely lost their lizard-like tail, and have not yet perfected their horny beak, but are only feeling their way as conquerors of the air. And as for the milk-givers, though we have met with them in small early forms, yet now for a time we lose sight of them again altogether.

It is the reptiles—the cold-blooded monster reptiles— which seem at this time to be carrying all before them. We find them everywhere—in the water, with paddles for swimming; in the air, with membranes for flying; on the land

193   Sauropsida.
194   Picture heading, Chap. III.

hopping or running on their hind feet. From small creatures not bigger than two feet high, to huge monsters thirty feet in height, feeding on the tops of trees which our giraffes and elephants could not reach, they fill the land; while flesh-eating reptiles, quite their match in size and strength, prey upon them as lions and tigers do upon the grass-feeders now.[195] This is no fancy picture, for in our museums, and especially in Professor Marsh's wonderful collection in Yale Museum in America, you may see the skeletons of these large reptiles, and build them up again in imagination as they stood in those ancient days when they looked down upon the primitive birds and tiny marsupials, little dreaming that their own race, then so powerful, would dwindle away, while these were to take possession in their stead.

And now in our series of changing scenes comes all at once that strange blank which we hope one day to fill up; and when we look again the large reptiles are gone, the birds are spreading far and wide, and we come upon those early and primitive forms of insect-eaters, gnawers, monkeys, grass-feeders, and large flesh-eaters, whose descendants, together with those of the earlier marsupials, are henceforward to spread over the earth. We need scarcely carry our pictures much farther. We have seen how, in these early times, the flesh-feeders and grass-feeders were far less perfectly fitted for their lives than they are now;[196] how the horse has only gradually acquired his elegant form; the stag his branching antlers; and the cat tribe their scissor-like teeth, powerful jaws, and muscular limbs; while the same history of gradual improvement applies to nearly all the many forms of milk-givers.

195   Picture heading, Chap. V.
196   Picture heading, Chap. IX.

But there is another kind of change which we must not forget, which has been going on all through this long history, namely, alterations in the level and shape of the continents and islands, as coasts have been worn away in some places and raised up or added to in others, so that different countries have been separated from or joined to each other. Thus Australia, now standing alone, with its curious animal life, must at some very distant time have been joined to the mainland of Asia, from which it received its low forms of milk-givers, and since then, having become separated from the great battlefield of the Eastern Continent, has been keeping for us, as it were in a natural isolated zoological garden, the strange primitive Platypus and Echidna, and Marsupials of all kinds and habits.

So too, Africa, no doubt for a long time cut off by a wide sea which prevented the larger and fiercer animals from entering it, harboured the large wingless ostriches, the gentle lemurs, the chattering monkeys, the scaly manis, and a whole host of insect-eaters; while South America, also standing alone, gave the sloths and armadilloes, the ant-bears, opossums, monkeys, rheas, and a number of other forms, the chance of establishing themselves firmly before stronger enemies came to molest them. These are only a few striking examples which help us to see how, if we could only trace them out, there are reasons to be found why each animal or group of animals now lives where we find it, and has escaped destruction in one part of the world when it has altogether disappeared in others.

*****

So, wandering hither and thither, the backboned family, and especially the milk-givers, took possession of plains and mountain ranges, of forests and valleys, of deserts and fertile regions. But still another question remains—How has it come to pass that large animals which once ranged all over Europe and Northern Asia,—mastodons, tusked tapirs, rhinoceroses, elephants, sabre-toothed tigers, cave-lions, and hippopotamuses in Europe,[197] gigantic sloths and llamas in North America, and even many huge forms in South America, have either been entirely destroyed or are represented now only by scattered groups here and there in southern lands? What put an end to the "reign of the milk-givers," and why have they too diminished on the earth as the large fish, the large newts, and the large lizards did before them?

To answer this question we must take up our history just before the scene at the head of our last chapter,[198] which the reader may have observed does not refer, as the others have done, to the animals in the chapter itself. Nevertheless it has its true place in the series, for it tells of a time when the great army of milk-givers had its difficulties and failures as well as all the other groups, only these came upon them not from other animals but from the influence of snow and ice.

For we know that gradually from the time of tropical Europe, when all the larger animals flourished in our country, a change was creeping very slowly and during long ages over the whole northern hemisphere. The climate grew colder and colder, the tropical plants and animals were driven back or died away, glaciers grew larger and snow deeper and more lasting, till large sheets of ice covered Norway and Sweden, the northern parts of Russia, Germany, England, Holland, and Belgium, and in America the whole of the country as far south as New York. Then was what geologists call the "Glacial Period;" and whether the whole country was buried

197　　Picture Heading, Chap. X.
198　　Chapter XI., Europe in the Age of Ice.

in ice, or large separate glaciers and thick coverings of snow filled the land, in either case the animals, large and small, must have had a bad time of it.

True, there were probably warmer intervals in this intense cold, when the more southern animals came and went, for we find bones of the hippopotamus, hyæna, and others buried between glacial beds in the south of England. But there is no doubt that at this time numbers of land animals must have perished, for in England alone, out of fifty-three known species which lived in warmer times, only twelve survived the great cold, while others were driven southwards never to return, and the descendants of others came back as new forms, only distantly related to those which had once covered the land.

Moreover, when the cold passed away and the country began again to be covered with oak and pine forests where animals might feed and flourish, we find that a new enemy had made his appearance. Man—active, thinking, tool-making man—had begun to take possession of the caves and holes of the rocks, making weapons out of large flints bound into handles of wood, and lighting fires by rubbing wood together, so as to protect himself from wild beasts and inclement weather.

In America and in England alike, as well as in Northern Africa, Asia Minor, and India, we know that man was living at this time among animals, many of them of species which have since become extinct, and with his rude weapons of jagged flint was conquering for himself a place in the world.

He must have had a hard struggle, for we find these flint implements now lying among the bones of hyænas, sabre-toothed tigers, cave-lions, cave-bears, rhinoceroses, elephants, and hippopotamuses, showing that it was in a land full of wild beasts that he had to make good his ground.

"By the swamp in the forest
The oak-branches groan,
As the savage primeval,
With russet hair thrown
O'er his huge naked limbs, swings his hatchet of stone.
"And now, hark! as he drives with
A last mighty swing,
The stone blade of the axe through
The oak's central ring,
From his blanched lips what screams of wild agony spring!
There's a rush through the fern-fronds,
A yell of affright,
And the Savage and Sabre-tooth
Close in fierce fight,
As the red sunset smoulders and blackens to night."[199]

Many and fierce these conflicts must have been, for the wild beasts were still strong and numerous, and man had not yet the skill and weapons which he has since acquired. But rough and savage though he may have been, he had powers which made him superior to all around him. For already he knew how to make and use weapons to defend himself, and how to cover himself at least with skins as protection from cold and damp. Moreover, he had a brain which could devise and invent, a memory which enabled him to accumulate experience, and a strong power of sympathy which made him a highly social being, combining with others in the struggle for life.

And so from that early time till now, man, the last and greatest winner in life's race, has been taking possession of the earth. With more and more powerful weapons he has fought against the wild beasts in their native haunts; and by clearing away the large forests, cutting up the broad prairies and pastures, and cultivating the land, he has turned them

---

199   From "A Legend of a Stone Axe," a clever and suggestive poem in the *New Quarterly*, April 1879. The text is slightly altered.

out of their old feeding grounds, till now we must go to the
centre of Africa, the wild parts of Asia, or the boundless
forests of South America, to visit in their homes the large
wild animals of the great army of milk-givers.

* * * * *

Since, therefore, these forms are growing rarer every
century, and some of them, such as the Dodo, Epyornis,
and Moa among birds, and the northern sea-cow or Rhy-
tina among milk-givers, have already disappeared since the
times of history, we must endeavour, before others are gone
for ever, to study their structure and their habits. For we
are fast learning that it is only by catching at these links
in nature's chain that we can hope to unravel the history of
life upon the earth.

At one time naturalists never even thought that there
was anything to unravel, for they looked upon the animal
kingdom as upon a building put together brick by brick, each
in its place from the beginning. To them, therefore, the fact
that a fish's fin, a bird's wing, a horse's leg, a man's arm and
hand, and the flipper of a whale, were all somewhat akin,
had no other meaning than that they seemed to have been
formed upon the same plan; and when it became certain
that different kinds of animals had appeared from time to
time upon the earth, the naturalists of fifty years ago could
have no grander conception than that new creatures were
separately made (they scarcely asked themselves how) and
put into the world as they were wanted.

But a higher and better explanation was soon to be found,
for there was growing up among us the greatest naturalist
and thinker of our day, that patient lover and searcher after
truth, Charles Darwin, whose genius and earnest labours
opened our eyes gradually to a conception so deep, so true,

and so grand, that side by side with it the idea of making an animal from time to time, as a sculptor makes a model of clay, seems too weak and paltry ever to have been attributed to an Almighty Power.

By means of the facts collected by our great countryman and the careful conclusions which he drew from them, we have learned to see that there has been a gradual unfolding of life upon the globe, just as a plant unfolds first the seed-leaves, then the stem, then the leaves, then the bud, the flower, and the fruit; so that though each part has its own beauties and its own appointed work, we cannot say that any stands alone, or could exist without the whole. Surely then Natural History acquires quite a new charm for us when we see that our task is to study among living forms, and among the remains of those that are gone, what has been the education and the development of all the different branches, so as to lead to the greatest amount of widespreading life upon the globe, each having its own duty to perform. With the great thought before us that every bone, every hair, every small peculiarity, every tint of colour, has its meaning, and has, or has had, its use in the life of each animal or those that have gone before it, a lifelong study even can never weary us in thus tracing out the working of Nature's laws, which are but the expression to us of the mind of the great Creator.

When we once realise that whether in attacking or avoiding an enemy it is in most cases a great advantage to all animals to be hidden from view, and that each creature has arrived at this advantage by slow inheritance, so that their colours often exactly answer the purpose, how wonderful becomes the gray tint of the slug, the imitation of bark in the wings of the buff-tip moth, the green and brown hues of the eatable caterpillars, the white coat of the polar bear, and the changing colour of the arctic fox, the ermine, and the ptarmigan, as winter comes on! And when, on the other hand, we find badly-tasting creatures such as ladybirds and some butterflies, or stinging animals like bees and wasps, having bright colours, because it is an actual advantage to them to be known and avoided, we see that in studying colour alone we might spend a lifetime learning how the winners in life's race are those best fitted for the circumstances under which they live, so that in ever-changing variety the most beautifully-adapted forms flourish and multiply.

Then if we turn to the skeleton and the less conspicuous framework of the body, the flippers of the whale, the manatee, or the seal, doing the work of a fish's fin and yet having the bones of a hand and arm, reveal a whole history to us when we have once learned the secret that in the attempt to increase and multiply no device is left untried by any group of animals, and so every possible advantage is turned to account.

Next, the wonderful instincts taught by long experience give us a whole field of study. We see how frogs and reptiles, and even higher animals such as marmots, squirrels, shrews and bears, escape the cold and scarcity of food in winter by burying themselves in mud, or in holes of trees or caves of the earth till spring returns; and while we find alligators burying themselves in cold weather in America, we find crocodiles, on the contrary, taking their sleep in the hot dry weather in Egypt because then is their time of scarcity.

Then we learn that the birds avoid this difficulty of change of climate in quite another manner. They with their power of flight have learned to migrate, sometimes for short distances, sometimes for more than a thousand miles, so that they bring up their young ones in the cool north in summer, when caterpillars and soft young insects are at hand for their prey, and lead them in the winter to the sunny south where food and shelter in green trees are always to be found. So long indeed has this instinct of migration been at work, that often we are quite baffled in trying to understand why they take this or that particular route for their flight, because probably, when the first stragglers chose it, even the areas of land and water were not divided as now, so that we must study the whole history of the changing geography of the earth to understand the yearly route of the swallow or the stork.

And last but not least, when we look upon the whole animal creation as the result of the long working out of nature's laws as laid down from the first by the Great Power of the Universe, what new pleasure we find in every sign of intelligence, affection, and devotion in the lower creatures! For these show that the difficulties and dangers of animal life have not only led to wonderfully-formed bodies, but also to higher and more sensitive natures; and that intelligence and love are often as useful weapons in fighting the battle of life as brute force and ferocity.

Even among the fish, which, as a rule, drop their eggs and leave them to their fate, we have exceptions in the nest-building sticklebacks and the snake-headed fish of Asia, which watch over and defend their fry till they are strong, in the pipe-fish where the fathers carry the young in a pouch, and in sharks which travel in pairs; while a pike has been known to watch for days at the spot where his mate was caught and taken away, and mackerel and herrings live in shoals and probably call to each other across the sea.

Among the other cold-blooded animals—the frogs, newts, and reptiles—it is true we find less show of feeling, but we

must remember that these are only poor remaining frag-
ments of large groups which have disappeared from the
earth. Even among the amphibia however a tame toad will
become attached to one person; while among reptiles, lizards
are full of intelligence and affection, and snakes are well-
known for their fondness for their owners. The case of the
snake which died by its master's side when he fell down
insensible,[200] if it can be relied upon, would show that even
cold-blooded animals have tender hearts.

Yet these are all instances of affection of lower animals
to man. We must turn to the birds, that group which has
gone on increasing in strength and numbers down to our
day, to find that tender devotion which watches over the
helpless nursling, defends the young at the risk of life, nay,
like the peewit with the dragging wing, will even run in
the face of death to lure the cruel destroyer away from the
hidden nest. Natural history teems with examples of birds
faithful to each other and pining even till death for the loss
of a mate; while many birds, such as rooks, starlings, wild
geese, swans, and cranes, not only live in companies and
exact obedience from their members, but even set sentinels
to watch, the duties of the office being faithfully fulfilled.

Then again it is to the higher animals, those nearer to
ourselves, that we must look for the truest affection, and the
strongest proofs of that obedience and sympathy which lead
them to unite and so become strong in the face of danger.
Among the beasts of prey it is true that, except the wolves
and jackals, none herd together; but family love is strong
and true. No tiger is so dangerous as is the mother tigress
if any one approaches her young ones, or the lioness whose
cubs are attacked, and in our own homes we all know the

200   *Animal Intelligence*, Romanes, p. 261.

tenderness and devotion of a cat to her kittens. Nevertheless, these animals have very little social feeling; theirs are
the narrower virtues of courage and fidelity to home, and to
the duty of providing food for wife and children. It is among
the gentler vegetable-feeders,—the antelopes and gazelles,
the buffaloes, horses, elephants, and monkeys,—that we
find the instinct of herding together for protection, and with
this the consciousness of the duty of obedience and fidelity
to the herd and to one another.

It is easy to see how this was necessary to protect these
feebler animals from the attacks of their ferocious neighbours,
and also what an advantage they had when they had once
learned to set sentinels who understood the duty of watching
while others fed, as in the case of the chamois and seals, of
obeying the signal of a leader like the young baboons on the
march, or of putting the mothers and children in the centre
for protection, as horses and buffaloes do.

And there is a real significance in this gradual education
in duty to others which we must not overlook, for it shows
that one of the laws of life which is as strong, if not stronger, than the law of force and selfishness, *is that of mutual
help and dependence.* Many good people have shrunk from
the idea that we owe the beautiful diversity of animal life on
our earth to the struggle for existence, or to the necessity
that the best fitted should live, and the feeblest and least
protected must die. They have felt that this makes life a
cruelty, and the world a battlefield. This is true to a certain
extent, for who will deny that in every life there is pain and
suffering and struggle? But with this there is also love and
gentleness, devotion and sacrifice for others, tender motherly
and fatherly affection, true friendship, and a pleasure which
consists in making others happy.

This we might have thought was a gift only to ourselves—
an exception only found in the human race; now we see that
it has been gradually developing throughout the whole animal
world, and that the love of fathers and mothers for their
young is one of the first and greatest weapons in fighting
life's battle. So we learn that after all, the struggle is not
entirely one of cruelty or ferocity, but that the higher the
animal life becomes, the more important is family love and
the sense of affection for others, so that at last a fierce beast
of prey with strength and sharp tools at his command, is
foiled in attacking a weak young calf, because the elders of
the herd gather round him, and the destroyer is kept at bay.

Surely then we have here a proof that, after all, the
highest and most successful education which Life has given
her children to fit them for winning the race is that "unity
is strength;" while the law of love and duty beginning with
parent and child and the ties of home life, and developing into
the mutual affection of social animals, has been throughout
a golden thread, strengthened by constant use in contending
with the fiercer and more lawless instincts.

So it becomes evident that the beautiful virtue of self-
devotion, one of the highest man can practise, has its roots
in the very existence of life upon the earth. It may appear
dimly at first,—it may take a hard mechanical form in such
lowly creatures as insects, where we saw the bees and ants
sacrificing all tender feelings to the good of the community.
But in the backboned family it exists from the very first as
the tender love of mother for child, of the father for his mate
and her young ones, and so upwards to the defence of the
tender ones of the herd by the strong and well-armed elders,
till it has found its highest development in man himself.

Thus we arrive at the greatest and most important lesson

that the study of nature affords us. It is interesting, most interesting, to trace the gradual evolution of numberless different forms, and see how each has become fitted for the life it has to live. It gives us courage to struggle on under difficulties when we see how patiently the lower animals meet the dangers and anxieties of their lives, and conquer or die in the struggle for existence. But far beyond all these is the great moral lesson taught at every step in the history of the development of the animal world, that amidst toil and suffering, struggle and death, the supreme law of life is the law of Self-devotion and Love.

www.ingramcontent.com/pod-product-compliance
Lightning Source LLC
Chambersburg PA
CBHW030452210326
41597CB00013B/638